Notes on Ore Sampling in Mines

by A.W. Warwick

with an introduction by Kerby Jackson

Introduction

It has been years since A.W. Warwick released his important publication "Notes on Ore Sampling". First released in 1903, this important volume has now been out of print for this days and has been unavailable to the mining community since those days, with the exception of expensive original collector's copies and poorly produced digital editions.

It has often been said that "*gold is where you find it*", but even beginning prospectors understand that their chances for finding something of value in the earth or in the streams of the Golden West are dramatically increased by going back to those places where gold and other minerals were once mined by our forerunners. Despite this, much of the contemporary information on local mining history that is currently available is mostly a result of mere local folklore and persistent rumors of major strikes, the details and facts of which, have long been distorted. Long gone are the old timers and with them, the days of first hand knowledge of the mines of the area and how they operated. Also long gone are most of their notes, their assay reports, their mine maps and personal scrapbooks, along with most of the surveys and reports that were performed for them by private and government geologists. Even published books such as this one are often retired to the local landfill or backyard burn pile by the descendents of those old timers and disappear at an alarming rate. Despite the fact that we live in the so-called "Information Age" where information is supposedly only the push of a button on a keyboard away, true insight into mining properties remains illusive and hard to come by, even to those of us who seek out this sort of information as if our lives depend upon it. Without this type of information readily available to the average independent miner, there is little hope that our metal mining industry will ever recover.

This important volume and others like it, are being presented in their entirety again, in the hope that the average prospector will no longer stumble through the overgrown hills and the tailing strewn creeks without being well informed enough to have a chance to succeed at his ventures.

Kerby Jackson
Josephine County, Oregon
October 2015

NOTES ON SAMPLING

INTRODUCTION.

During the year 1902 a number of articles were published in the Denver Mining Reporter under the caption of "Notes on Sampling." In response to a somewhat widespread demand these notes are republished with a few slight alterations and additions.

These notes came to be written under somewhat interesting conditions. The late Mr. Henry A. Vezin of Denver had for a great many years been in the habit of making numerous studies upon all engineering problems relating to mining and metallurgy. Out of these studies grew a great mass of papers and letters, all carefully docketed and indexed according to subject matter. Among the most numerous of his papers were theses—facetiously called by Mr. Vezin, "tape-worms—upon sampling. Mr. Vezin, with his usual broad-minded liberality gave the present writer the most complete access to his wonderful collection of papers. No one who has ever had the privilege of so doing could fail to be impressed by the thoroughness and keenness of Mr. Vezin's investigations upon any subject he studied. It seemed a thousand pities that such valuable works were not brought before the notice of the mining world. It was suggested to Mr. Vezin that these papers should be edited and published. He gave cordial assent, but with characteristic modesty he insisted that the notes should be published anonymously. This was reluctantly done.

The original intention was to publish a few articles upon the difficult and complex subject of ore sampling, the articles to be composed of extracts from Mr. Vezin's papers. In order, however, to make the matter more connected, a great many new studies were made, and many of the papers were re-written.

Mr. Vezin was known as one of the greatest authorities on sampling and it seems to the writer that it would have been a distinct loss to mining and metallurgical engineers if Mr. Vezin's views, and those indorsed by him, were not given to the world.

At the time of Mr. Vezin's death, on December 27. 1902, a number of plans were being made to continue these notes, especially along the line of design of large sampling works. In the midst of these plans, however, Mr. Vezin was stricken down by heart failure and the present writer feels too much diffidence to carry them out alone.

During the last few years of Mr. Vezin's life, the writer occupied a position of peculiar intimacy with him. With tireless mental activity was combined that rarest of all characteristics, a desire to help the younger men who would have to carry on his work when he haid laid it aside. His notes and studies were at the service of all who wished to make legitimate use of them.

It is with poignant sorrow that the writer passes the following pages through the press without Mr. Vezin's keen and kindly criticism. Within the following pages there will doubtless be found errors that never would have occurred had Mr. Vezin been alive to correct and criticise them.

A. W. WARWICK.

CHAPTER I.

Few subjects have been so frequently discussed as ore sampling. The discussion, whilst apparently endless, is very necessary. It seems impossible to fix in the mind of the mining public the very simple laws that govern the operation of sampling. No operation is easier to perform with certainty and accuracy, than that of sampling. Many intelligent men, however, seem to look upon sampling as a very uncertain and difficult operation. In fact, some time ago an ore buyer in Leadville insisted that it was impossible to get a close sample from rich ore, and that a species of compromise had necessarily to be accepted. That conclusion was not to be wondered at after seeing the way in which he sampled. One of the most glaring examples of bad sampling that ever came under my notice was recently performed by an engineer widely known on both sides of the Atlantic. In the report issued by that engineer, duplicate samples varied so enormously in value that it is inconceivable how any engineer could have accepted such results, much less have published them. The assays were made by two different assayers, each most reliable. The following are a few examples from the report referred to:

Example.	Assayer 1, ozs. gold	Assayer 2, ozs. gold.	Differences per ton.
1	10.08	5.36	$ 94.40
2	0.32	21.94	432.40
3	5.68	1.04	92.80
4	2.40	0.21	43.80
5	3.12	10.00	137.60
6	79.04	11.46	1,351.60

These are but a few of the extraordinary results obtained from two halves of the same pulp. This, be it remembered, of ore which it is customary to expect buyer and seller to agree within one dollar per ton. With greater differences the aid of an umpire is called in. Correct sampling is of more importance to-day than ever before, owing to the fact that miners are selling practically the whole of the output of their mines instead of milling or smelting their own ores. Thus in Colorado, with a mineral output aggregating more than $50,000,000 per annum, an error of two per cent. in the settlement would entail a loss of $1,000,000 to some one.

The subject is, therefore, of much importance, and apparently not yet well understood. The following notes on the subject have been therefore extracted from a mass of correspondence extending over the last ten years.

In sampling an ore pile one of two distinct modes may be used:

1. Hand sampling of which four distinct modes exist: (a), Coning and quartering; (b), split shovels; (c), alternate shovels; (d), riffling.

2. Machine sampling, which imitates one or another of the methods of hand sampling: Part of stream for the whole time. Whole stream part of the time.

Whatever method of sampling be adopted, it must be obvious that a distinct relationship must exist between the weight of the sample and the size of the ore particles. Thus, if the sample is to be at all representative, and the ore is as large as, say cocoanuts, then a large sample must be taken; if the ore is crushed to, say twenty mesh screen, a sample need only weigh but a few ounces to certainly represent correctly the whole mass of the original ore. How small the sample may be depends, also, upon the ore. It is far more difficult to sample an ore largely composed of barren material irregularly scattered through which is a very small amount of high grade material, than an ore more uniformly mixed or of greater homogeneity. The size of the sample made depends therefore not only on the size of the pieces of ore, but also upon the nature of the ore.

As a sample the following table is given. It was made many years ago for Gilpin county ores, carrying from one ounce to four ounces gold to the ton:

Diameter of largest pieces in in.	1-25	1-12	1-6	5-16	5-8	1 1-4	2 1-2
Minimum weight of sample lbs..	1-16	1-2	4	32	256	2,048	16,343

It will be noticed that the size of the sample increases as the cube of the size of the largest piece; thus with pieces of ore twice as large in one lot as another, a sample eight times as large will be taken.

When Mr. Philip Argall was managing the works of the Metallic Extraction Company at Cyanide, Colorado, he

---

*Originally he used two samplers, arranged for taking duplicate samples, each taking 1-16 of the stream of ore, for the purpose of checking the one sample against the other. With ore containing one or two ounces of gold to the ton, the samples checked so perfectly that he gave up the system of making the two samples; simply ran the material together over the two machines, thus giving an eighth at each division. The eighth was taken for safety, especially as the ore delivered was sometimes much richer.

reduced the ore, after being crushed to 1½-inch in diameter, by means of an automatic sampler to 1-8. This eighth which was allowed to accumulate on the crushing floor, was then passed through the same machinery, giving a resulting sample of 1-64 or 1 9-16 tons per 100 tons sampled.

pler took out ¼ as a sample, one scoopful every 1½ seconds; and this sample was passed to strong crushing rolls, which reduced it to ¼-inch in diameter. It was then elevated to another sampler, arranged to take out 1-10, one scoopful every 1½ seconds. The final sample was then

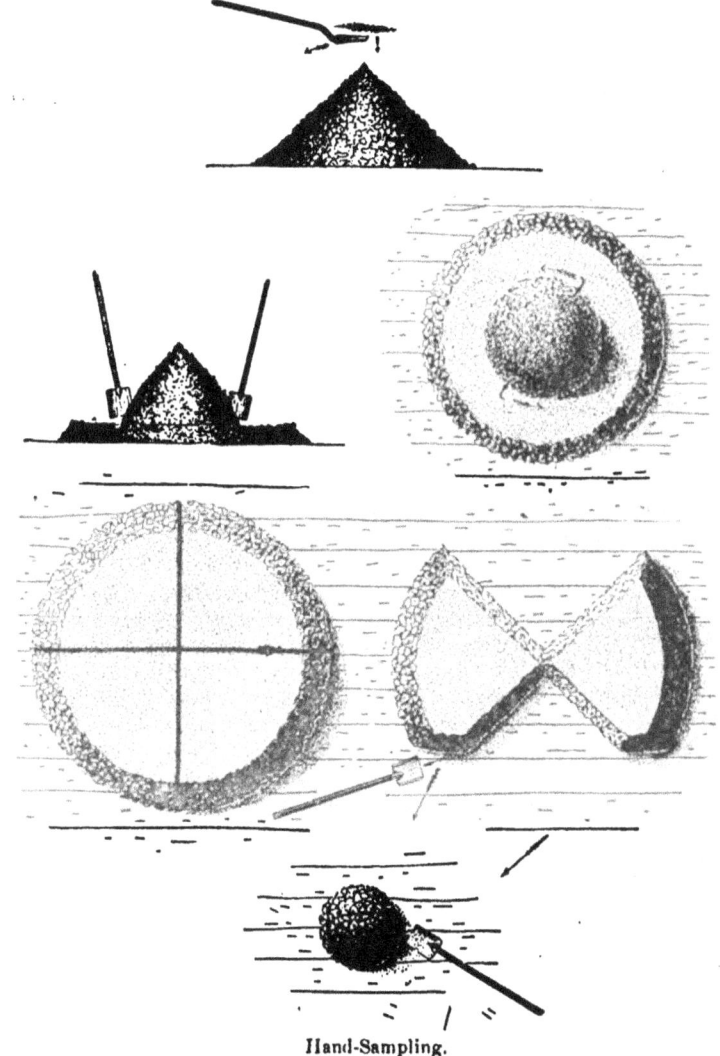

Hand-Sampling.

This was then crushed finer and divided down by taking alternate shovels, or by riffles. The results were perfectly satisfactory.*

In the extension of these works, built later, called mill No. 2, the ore was crushed as before to 1½ inches; the sam-

only 1-40 of the ore, or 2½ per cent., and was divided down and crushed finer in the usual way. By this method, a greater amount of the ore was crushed fine than would be approved of in smelting works, but, as all of it had finally to be crushed to thirty mesh, there was no objection to this.

The object of the smelter is to obtain a sample as representative as possible without crushing the ore so fine as to give trouble in the blast furnaces. According to the table already given, this is possible. Thus, 100 tons of a lot of ore is crushed to a two-inch ring. A correctly taken sample of this need not weigh more than about 6,500 pounds. This sample is then reduced to say one-inch guage, and a sample of 850 pounds will give a fair representative sample. This can be reduced to a ¼-inch mesh, and a sample of twenty-five pounds taken. By reducing this sample to a sixteen-mesh screen, a sample of ½ pound may be taken. Thus ½ pound, if the operations have been correctly performed, will represent accurately the original 100 tons. At the same time, the lot of ore will not have been broken down to a size that will give trouble in the furnaces.

Coning and Quartering.—This method of hand sampling, sanctified, but not sainted, has long held the field as the only reliable method of sampling and as the standard with which all other methods are compared. I object to the method for three reasons: First, it is extremely slow; second, it is expensive; third, it is not always exact, even if carried out with the utmost care. To the first two objections, but few practical men will demur. The third objection is, I know, very much like attacking a fundamental article of faith in theology, but the objection is nevertheless perfectly valid. In the old days of smelting and milling when hand sampling was in vogue, disagreements between buyer and seller were far more frequent than they are now, when machine sampling is almost universally used. The requirements of the ore seller are now very much more severe than formerly, and it is more than probable that this increase in strictness of requirements is due to the certainty and accuracy of machine sampling.

By the method of quartering, one-half of the original pile of ore is taken at each reduction. The thorough mixing of the ore is an absolutely necessary condition in quartering down ore. This mixing is supposed to be effected by coning. The cone is built up by men moving around the circumference of a circle and shovelling the ore upon the point of a cone formed by the angle of repose of the material falling vertically on one given point. The samplers—from four to eight men—move so as to always be diametrically opposite each other. The ore pile, from which the cone is made, is thus transferred to the cone, and the floor swept up quite clean; these sweepings are also thrown on to the top of the cone. A cone containing twenty tons of ore will be about seven feet high, and will have a base of about twelve to thirteen feet in diameter. The cone is then "pulled down" by the men walking around the pile and scraping the ore down from the apex of the cone toward the base. This operation is continued until the whole

mass is converted into a plaque of ore about eighteen feet in diameter and from twelve to fourteen inches thick. The workmen then take a long straight-edge and mark two lines at right angles to each other, and intersecting at the center of the circle. The plaque is thus divided into four quarters. The two opposite quarters are removed to the ore bins, the remaining two are mixed into a second cone, and the whole cycle of operations is repeated until it becomes necessary to recrush the ore. The sampling is thus continued until a parcel of twenty-five to thirty pounds is made. This is then crushed very fine, and the sample prepared for the assay office. The whole process is slow and laborious; two men and a shoveler being able to thus reduce down from twenty to twenty-five tons per shift at a cost of about 45 or 50 cents per ton. The sight of a gang of men sampling by the method of coning and quartering is very laughable. The men are cautioned to be careful and deliberate, and they carry out these orders most conscientiously.

Some further objections could be urged against coning and quartering. In the old days, when Professor N. P. Hill first came to establish his works at Black Hawk, he saw the many chances of errors in the "sacred method of coning and quartering." He saw that even with care, the point of the cone would travel away from where it was started, and in consequence one of the quarters would take out a large part of the center of the lower part of the pile. To correct this he used a rod, driven into the floor as a guide. Then, when the cone had been flattened down, in taking out the quarters, it was impossible to make a clean division, and a good deal of the coarse stuff from the quarters remaining would roll down into the quarters taken away. To avoid this error, he had plates of iron which would be pressed down into the ore, dividing the quarters by them. There comes still another objection, and that is, that when the men throw shovelful after shovelful upon the point of the cone, the way in which they hold their shovels may certainly interfere with the regular distribution of the stuff, and this is the great end of the coning. If the stuff could be thrown upon it from a solid stream, falling vertically, this objection would disappear. About the same time, without collusion with Professor Hill, I (the original maker of these notes) not knowing him, suggested dumping the whole into a cone ending in a cylindrical portion which was immediately over a cross that divided the stuff into four quarters. A plate, serving as a gate, closed the discharge of the cone. When filled, the gate was withdrawn quickly; the solid stream was divided perfectly by the cross. Two opposite quarters were taken as a sample, the two others rejected. Those that formed the sample could be gathered into a similar cone, treated in the same way, then over a third one, so that the final sample of the two opposite quarters was one-eighth of the original, my idea being that the ore would then have to be crushed finer before being cut down any more. The same machine might serve for the further cutting down after crushing, by having the crushed sample delivered by means of an ordinary cup elevator to the top cone, or be hoisted by means of a platform elevator. Or, but a single cone could be used, the two quarters forming the sample, received in cars or other suitable vessels, hoisted by means of a platform elevator dumped into the cone, and this operation repeated as often as might be considered safe. As one man could attend to all, the sampling would be so much safer against the chances of salting. This system could, of course, be applied to the original ore, that is to say, the halving done of the whole lot of ore. I never carried this out, but I had a small apparatus of this kind made for dividing down laboratory samples. I did not have the gate because pouring the crushed ore from the paper in which it was mixed, taking the direction of one of the edges of the cross, I could throw it into the cone quickly enough to secure a solid stream. The whole thing was cut down without a spatula, without sweeping up, and without the amount of manipulation which seems to be in favor in assay offices. I prefer to this arrangement the riffle sampler first introduced by Professor Hill.

CHAPTER II.

We have a very small opinion of the intelligence of those men who still persist in employing the method of sampling known as coning and quartering. We condemn it utterly as being costly, slow and by no means accurate. We do not, however, condemn all methods of hand sampling as being incorrect; on the contrary, we recognize the need of a rapid and accurate method. This, we believe, is to be found in the method of "alternate shovels." In the coning and quartering method, one only takes two large cuts. If the ore is not homogenous or thoroughly mixed when the two first cuts are made, the resulting half cannot possibly represent the whole mass of ore. It follows, then, that not even the most scrupulous care, or the most perfect method of cutting down this large sample, to the size necessary for the assay office, can possibly give a correct result. Mr. Vezin says: "In hand sampling, I adopted in 1875, or even earlier, the method of taking alternate shovelfuls, and in 1879 I introduced it in the smelting works of the Little Chief mine on Fryer hill, Leadville. Why the managers of the other works did not follow this good example, is a mystery to me. Perhaps it was due to their unconquerable aversion to any innovation."

This method is used, even at the present day, by those who are not progressive, and who are afraid to adopt automatic sampling, or even the simpler and more accurate method of taking alternate shovelfuls. The large number of cuts in sampling by taking alternate shovelfuls insures greater accuracy than in coning and quartering, and the labor involved is about one-fourth as great. Assuming that a shovelful of ore weighs ten pounds, and that a pile of ore of twenty tons is to be cut down, then the sample would contain 2,000 cuts, while in the Druid system, there would be but two cuts. If the ore is of such a nature that taking every third or fourth shovel for the sample is safe, then the sample would consist of 1,333⅓ or 1,000 cuts, respectively, against the two cuts in the Druid cone, and the cutting down would take place so much faster. The writer not inaptly compares the method of taking alternate shovelfuls of ore to cutting out alternate sections from a long narrow ribbon of ore passing before the sample men. It must be obvious to any man of ordinary intelligence that by such a method no part of the ore can escape being sampled. Whereas, when a huge slice of ore is taken from the flattened out cone, in which it is practically impossible to get anything like a perfect mixing, and hence a true representative sample. By the method of taking alternate shovelfuls no mixing is needed.

Automatic samplers have suffered considerably from having the work done by such machines checked by the barbarous method of coning and quartering. It is perfectly ridiculous to attempt to check the work done by an automatic sampler by the samples obtained by such a faulty method as coning and quartering. It is perfectly inconceivable how authors of elaborate and highly mathematical articles on sampling, can publish so-called "checks" of automatic sampling by the method of coning and quartering.

Of course, it may be necessary to convince the "practical" man that a machine gives a correct sample, since it may be impossible to convince him by a mathematical demonstration. But it is quite delusive to attempt to check a machine by a notoriously inaccurate method. What would be thought of a man who insisted upon testing an accurate and highly finished chronometer by means of a dollar watch? Would he not be considered crazy? Yet, it is just as ridiculous to test the accuracy of a new sampling method by the coning and quartering method. The writers, however, state that whilst the coning and quartering is to be condemned, yet the method of alternate shovelfuls is thoroughly to be relied upon, and it is by this latter method that he would test any machine or method.

Besides accuracy, other advantages are claimed for the method of alternate shovelfuls. In any well organized business, speed is recognized to be of the utmost importance. It should be the aim of the smelter or millman to obtain a sample as quickly as possible, and thus to be able to make a prompt settlement with the sellers. The ore is then transferred from the sampling department and into the beds or mill bins at the earliest possible moment. All practical men recognize this. By the method of coning and quartering, a lot of ore may lie around awaiting sampling for several days. Even after the men get to work, some time must elapse before a sample is obtained. Again,

in the smelters and mills employing the coning and quartering method, the sampling works form by no means the most insignificant buildings of the establishment.

With the method of alternate shovelfuls, not only is the cost at least one-fourth that of the older method, but it is also quite three or four times as fast, and thus fulfills the conditions laid down above. Nor is this all, for the buildings required by the sampling department will be smaller in size, and fewer in number.

To summarize the advantages of the method of alternate shovelfuls over the method of coning and quartering.

1. It is more reliable and more accurate.
2. It is cheaper in operation.
3. It is quicker.
4. Less space is required.
5. The investment of capital required is much smaller, with the consequent saving in interest on capital.

Yet in spite of the advantages, our metallurgists still carry on the laborious method of coning and quartering, priding themselves upon the up-to-date character of their work, and the "conservative" and business-like manner in which they direct it.

CHAPTER III.

The varied opinions expressed as to the value of different methods of sampling, and the merits and demerits of various machines, seem beyond all reasonable explanation. It seems to be a case of Quot homines tot Sententiæ—so many men, so many minds. It is unsettling, to say the very least, when a man of reputed ability and intelligence positively states that a machine or process of sampling is unreliable; he has tried it and found it inaccurate. On this question, Mr. Vezin, in one of his letters to a business friend, says: "One of the ablest metallurgists of the country, a man of remarkable ingenuity, and especially able in adapting means to ends, assured me that he had found the ordinary riffle sampler unsafe. That is to say, that by halving the sample successively by one of these, the two final samples did not agree. I could not understand how this was possible until he explained to me that for the purpose of making the test as severe as possible, he had always taken the portion that passed through, for the one sample, and the portion that was retained in the riffles for the other. I did not see that that could make the slightest difference until he explained that he piled the stuff used, higher than the top of the riffles, so that, when he lifted the sampler, the coarse particles which laid over the riffles

would roll down into the open spaces. The result was that, in some samples, the part retained in the riffles was richer, in others poorer, as there are some ores in which the fine contains much less than the coarse. When a man of intelligence will use a machine as it is not intended that it should be used, there is no end to the objections which may be made to any device that is suggested for any purpose."

Car Samples.—The statement of the above experience will certainly explain why so many groundless objections are made to various forms of sampling machines, or methods of sampling. It will also explain why some people yet use methods which are admittedly incorrect. They will use a good process so badly that the results obtained are useless, and therefore will condemn it utterly. The only recourse open to them is to go back to the old process which they follow out without the slightest deviation from the approved method of using it. It is the old idea of obtaining better satisfaction from an old inaccurate method carefully applied, than from a new and accurate method, improperly applied. An incident which called forth a long series of letters was as follows: An able metallurgist. managing a large cyanide mill to treat Cripple Creek ores, "checked" the results obtained from a very perfect form of mechanical sampler. He was not satisfied, and was in a state of fear and trembling lest he should be paying a dollar or two more per ton for the ore than it contained. A perfectly ruinous state of affairs. He then appealed to Mr. Vezin for assistance in his dilemma. It developed that the method of "checking" the results was simply to take a scoopful from the top of every car taken to the charging vat; a most strange and inexplicable phenomenon. "Why should not the results check?" asked the metallurgist. Of course, he would not expect the samples to agree absolutely from day to day, but at the end of every month he would expect to have a commercially accurate agreement. Sometimes one method of sampling should give a little higher result, sometimes the other.

The reply was to the effect that the method of taking a car sample was very imperfect. That a small sample weighing less than a pound, taken from the top of a car, cannot represent the true bulk of the ore in the car. That such a sample violated the first principle of the science of sampling which may be written "that in order to properly obtain a sample of ore, it is necessary to take the same quantity of ore frequently, or in as many places as possible, and at regular intervals." The metallurgist appeared to believe that mere crushing of the ore would so intimately mix it that a small quantity taken from the top of the car would approximately represent the whole of the car. The engineer vigorously denied this, stating that nothing is more difficult to do than to mix large quantities of ore so as to obtain a truly homogeneous mass. It was a fallacy to suppose that the mere running of ore through a mill would give a thorough mixing. In these two positions, the writer is undoubtedly correct, and it is but fair to quote once more that it is ridiculous to attempt to compare the work done by any sampling machine by taking so called car samples. Finally the engineer advised the metallurgist to take the ore rejected by the sampler and sample by alternate shovels. Then if, in a fifty-ton plant, it was found that the automatic sampler made a loss to the mill of from $50 to $100 per day to throw the machine over the dump.

Yielding, as all must, to the engineer's position, there is a reiteration in the metallurgist's letters to the effect that the car samples always averaged lower than the automatic sample. This is a point that appeals to us strongly, as it brings back an experience of our own. We may be forgiven, therefore, if we interject a few words of our own. In a small leaching mill, we took an automatic sample of the ore as it passed through the mill. We then took a "car sample" as the ore went to the tanks. We found that from month to month, the car samples ran less than our automatic sample, but on taking stock at the end of the year, we had extracted 6,000 ounces of silver more than were shown between our tailings samples and the car samples, but that we had 200 ounces less than the difference between our automatic samples and the tailings samples. This was repeated the next year, and it seems to us that some explanation is desirable. We offer the following:

It hardly seems necessary to point out that it is impossible to pile finely crushed dry ore containing particles of free gold or heavy mineral, so that a perfect mixing may be obtained. In the first place it is quite possible that the finely crushed particles of ore in falling into the bin, may separate; in the second, the vibration of the mill may accentuate this action when the ore is in the bin; thirdly, exactly the same events may occur whilst the ore is being discharged into the car. The final result will be, of course, that the material in the car will be unevenly distributed, and the specifically lightest and lowest grade material will be at the top. Hence, car samples of finely crushed material, are almost invariably low. The segregation of the high grade particles in a dry and crushed ore is well known, and in the coning and quartering system, this is guarded against by damping the ore during the final samplings. Yet this able and intelligent metallurgist expected an agreement between car samples and an automatic machine of great accuracy. Had the samples agreed, the perplexed and distraught manager would have been perfectly satisfied that at last he had obtained a machine of proven value. Had the samples agreed, then indeed there might have been some cause for doubt as to the accuracy of the machine-made sample. But it is interesting to learn that the worries of the manager became assuaged when he discovered that the actual returns yielded by the mill were in excess of those indicated by assays.

CHAPTER IV.

Quartering Shovels.—An appliance known as a quartering shovel was patented by Mr. D. W. Brunton in 1891. The object of this shovel was to put into the hands of the workman an instrument which would allow him to rapidly and accurately quarter down a pile of ore without recourse to the system of coning. These objects were not attained, although it is stated by Professor Hofman[*] that results

one being closed at the back, whilst the others are left open. The center compartment is about two and one-half inches wide. In using this shovel it is stuck into the heap and filled. The workman turns and gives a sharp rotary motion to the right, and in so doing the ore in the outside compartments will be jerked out of the shovel, and will leave the center compartment full, or partially so, of ore.

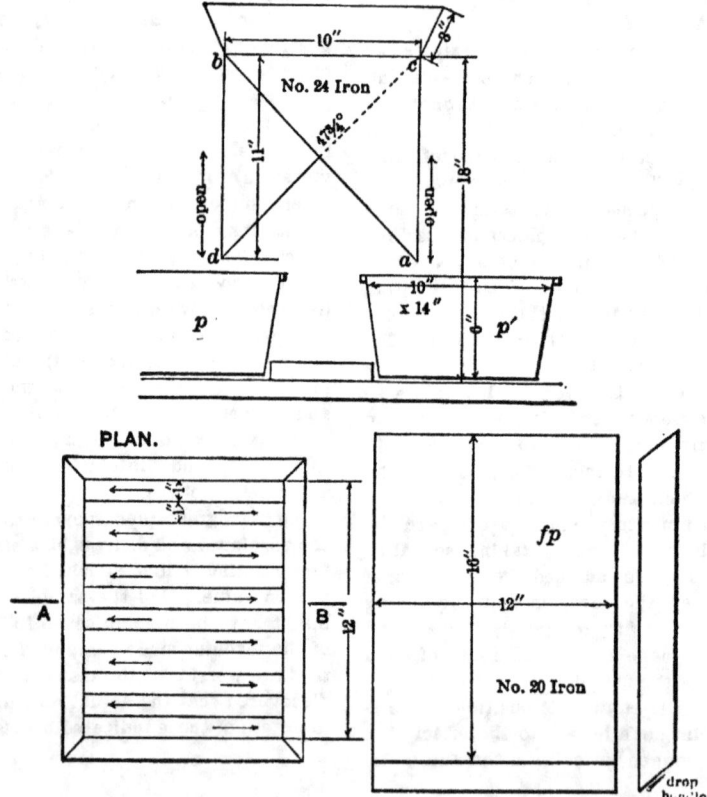

PLAN.

were obtained which checked well with those obtained from what our engineer calls the coning and quartering system. As Professor Hofman does not state what the character of the material was, we of course cannot say whether these tests were really satisfactory or not.

The shovel is about ten inches wide, and has turned-up sides. It is divided into three compartments, the center

This portion forms the reduced sample.

From this description it will be readily understood that the instrument cannot give a correct sample from coarsely crushed ore, in which the fines contain very high values. In sampling such an ore the large pieces on the top of the center compartment will almost certainly fall over into the rejected portion. Moreover, in filling the shovel more fines are apt to get into the center compartment than in the outside compartments. Thus the sample finally obtained carries more fines than the average of the ore, and as these fines are richer or poorer than the coarse, so will the sample be richer or poorer. The quartering shovel is an unreliable sampler. Moreover, it offers no advantage in point of speed over the method of alternate shovelfuls, and is not nearly as reliable. This sampler was evidently an attempt to improve upon the old single riffle sampler, which was sometimes attached to a handle. It was used in the following manner: One man held the riffle, while another man with an ordinary shovel threw the ore which was being sampled over the riffles. After each shovelful thrown

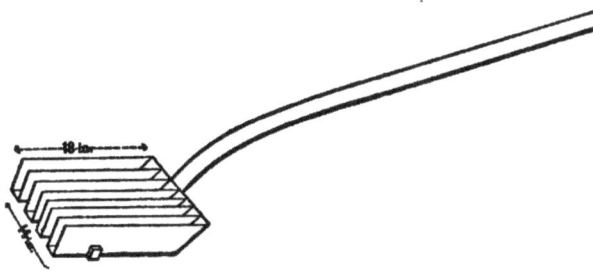

over the riffle, the man holding the latter turned to one side and emptied the riffle. This portion formed the sample. It hardly needs to be pointed out that this method of sampling was abominably slow and inaccurate.

Split Shovels.—The so-called split shovel is an instrument not unlike the quartering shovel. The split shovel resembles a long fork with the prongs made in the shape of deep channels. To this instrument the same objections urged against the quartering shovel hold. In unskillful or careless hands it is even more unreliable, as can be readily seen from the following description of the method of use: The split shovel is placed on the ground by one of the

sampling men, and another man, who faces the first, with a square-end shovel, delivers the ore from the pile over the entire width of the split shovel. The troughs or channels in the splitter are thus filled, whilst a certain proportion falls between and is rejected. Should the workmen be careless and allow the channels to become filled at any point and still continue to pour ore over the same point, then a certain amount of concentration will be effected, and the sample so obtained will be worthless. This applies more especially when the ore is formed of both coarse and fines.

The method of sampling by split shovels is very slow, certainly no faster than coning and quartering. It requires also very conscientious and careful workmen. This method is, therefore, of no value. It appears to be a backward improvement on the quartering shovel, that is to say, it is equally as inaccurate, and requires two appliances and double the amount of labor to obtain these results.

Riffle Samplers.—After a sampler has been reduced down to a certain size, the methods of sampling already mentioned are inapplicable, except, of course, the Druidical methods of coning and quartering, which, being inaccurate, should not be resorted to. Also, when by mechanical sampling the ore is reduced to a small amount, some other method of sampling must be resorted to. In such finishing processes nothing is better than the Jones sampler. This may be described as a nest of riffles with bottoms inclining at a considerable angle, each alternate riffle or channel having the bottom incline in an opposite direction. It will be obvious that by delivering the ore in a thin stream over the riffles, that two portions will be obtained, either of which can form the sample, since each should be equal both in grade and quantity. To further reduce the sample it must be run through the sampler again. The apparatus is very simple, is easily cleaned and gives rapidly a correct sample, if the ore is finely crushed.

———————

*Metallurgy of Lead, p. 51.

CHAPTER V.

Quartering Shovel.—The statement has been criticised that in pushing a quartering shovel into a pile of ore that the centre compartment will necessarily take in more fines than the outer compartments. The reason of this might be somewhat obscure, and, possibly, this point should have been more fully considered. It may be first stated that the fact noted is a fact, and not a mere supposition. A sample so obtained was subjected to a screen analysis. The ore had been crushed to a three-inch ring approximately. The sample contained forty-seven per cent. of the material that would go through a half-inch mesh, tne rejected portion only forty-one and one-half per cent. We explain the fact this way: Had the widths of the compartments been the same, the tendency for the coarse lumps to go into the different compartments would also be the same, but with one narrow opening and two wide openings, one on each side, there is a greater tendency for the lumps to jam in the entrance of the smaller compartment, and be forced into the wider compartments. To use a homely illustration, two men can go abreast through a wide door more easily than through a narrow door, and will be less likely to strike the jamb; so with the shovel.

Scoop Sampler.—It frequently happens that approximately correct samples are required quickly or without going to the trouble of handling the whole bulk. For example, crushed ore in sacks, loose piles of tailings, piles of concentrates, crushed ore, etc. For such work, the scoop sampler is very useful and gives approximately correct samples if **properly used**. This instrument resembles in form the old-fashioned cheese or butter taster, and is frequently made out of a piece of half-inch pipe which is cut in two by longitudinal slits. A handle of wood is attached to one end of such a piece of pipe, and the other end is turned up so as to close the channel. To use this instrument, the sampler is twisted like an augur into the pile. The ore caught in the channel forms the sample. In sampling sacks, the latter have to be tilted to one side so that the ore may stay in the scoop. In sampling piles of ore, strict attention must be paid to the fundamental law of sampling:

"That in order to properly obtain a sample of ore, it is necessary to take the sample frequently, or in as many places as possible, and to take the same quantity each time, at regular intervals."

We remember seeing, on one occasion, an alleged mining engineer sample a cone-shaped pile of ore which had been made up of different lots of crushed ore, each varying greatly in richness. The "sample" was obtained by walking around the pile, which contained about eight tons, and sticking a scoop sampler into the pile at irregular intervals. The sampler he had used he had brought with him for the purpose of sampling the pile. His smug look of self-satisfaction was very funny, and he remarked, "Well, I guess, that's a pretty fair sample." On his attention being called to the fact that the sample, in our opinion, was worthless, he positively asserted "That it couldn't help being a fair sample," and we fell below zero in his estimation because of our scepticism. The first law of sampling he couldn't or wouldn't understand. To prove that the sample obtained was worthless, the pile was shoveled over once, and thus partially mixed, and spread out into a flat layer about eighteen inches thick. The scoop sampler was used as before prodding the pile, at a slight angle from the vertical, at intervals of about a foot. The samples were assayed and were found to contain respectively (1) that taken by the engineer, thirty-two ounces in silver; (2) our sample, nineteen ounces, or a difference of thirteen ounces per ton. What happened in the first case was that the richer lots of ore were piled first, the lower grade ores last. By sampling a cone-shaped pile of ore, the amount of sample taken from the centre is greater in proportion than the amount taken from the outside; further, if the scoop reaches fully to the centre of the pile at each probe, the disproportion is much increased. Therefore, the "sample" taken by the self-satisfied M. E. was much richer than it should have been. He had violated the first law of sampling by taking more ore from one place than another, and not at regular intervals, as can be seen from the illustration. The simplest operation and the simplest instrument requires intelligence in the operator, and some application of that intelligence to the case he may have under consideration. As pointed out, time and again by our engineer, no machine or appliance exists that may not be misused.

So, too, with the scoop sampler. For certain operations it is reliable, such as for making rapid approximations, but, for certain other operations, such as sampling ore for sale, it is not close enough. But in the cases where it is applicable, some attention must be paid to the fundamental law of sampling.

Mixing.—Now we come to the great stumbling block,

for many engineers, of mixing. Many engineers feel themselves perfectly competent to pass on such matters as sampling even if they never have given the subject anything but the most cursory thought. Certain shibboleths have been set up, and by these all processes are promptly disposed of. Psittaceous arguments uttered with an air of profound wisdom have often an effect far beyond their value. Amongst these sayings none have been more overworked than the one which states that: "A good sampler must be a good mixer." So firmly has this shibboleth been fixed in the mind of the mining public that it seems necessary to point out its utter fallacy. In the first place, an attempt at mixing is made with the idea that the first law of sampling, so frequently referred to, might be violated, or, at all events, that the law need not be so rigidly applied. To the man who solemnly utters the shibboleth— "that to be efficient, a sampler must be a good mixer"—it may be pointed out that, unless a perfectly gigantic machine were built, only a few pounds of ore are ever in the sampling machine at any one time. It follows that only a few pounds could be mixed, and how this would advance the mixing of the whole parcel which is being sampled, is altogether obscure. Let us start with a bin full of ore, let us have an ore which is not homogeneous, say a Cripple Creek ore with the fines quite rich, and the coarse much poorer. Or that a car load of ore is made up of two lots, of different grades, which are to be treated as one. Any one who has had any experience in milling will know how obstinately each different grade of ore will remain by itself, or go through the mill by itself, and how small is the tendency of different qualities or grades of ore to mix with each other. Now, we have a bin of ore, the strata of which are of different grades or qualities. To make the case more pointed, suppose the lower stratum of ore is of lower grade than the top stratum. Very well, the lower stratum is drawn off first and passes through a perfect "mixer and sampler," and it may well be asked how can that effect a mixing of the ore contained in the bin? It is, of course, impossible. This fable as to mixing being a necessity in sampling possibly arose from the need of a thorough mixing, or an even distribution of the different grades, in the Druidical system. It is one of the rules of logic that comparisons cannot be drawn betwen unlike things, and nothing can be more unlike than machine sampling and the old happy go easy methods of the Druids. Yes, but, says the captious critic, you cannot "help" but get a better sample if the machine is a good mixer, and moreover the ore cannot help but get mixed whilst going through the mill. More shibboleths, more psittaceous arguments. Let us give some examples to prove that mixing of ores is a very difficult thing to do. Our engineer points out two facts, first, that in a sample prepared for the assayer, any number of "roll-ings" will not give a perfect mixing, so that the assays made from different parts of the pulp will agree within the assay error; second that in the operation of making a uniform cotton yarn for spinning, a good uniform thread can only be obtained by folding the cotton over on itself many thousands of times. How difficult then is the mixing of coarsely crushed, non-homogeneous ores.

Now let us give two more examples, both from milling practice. Until the last few years every stamp mill erected in California had a grizzly to remove the fines from the ore before it was sent to the rock breaker. The idea being to relieve the rock breaker of unnecessary work. The grizzly was set over the ore bin in such a position that the

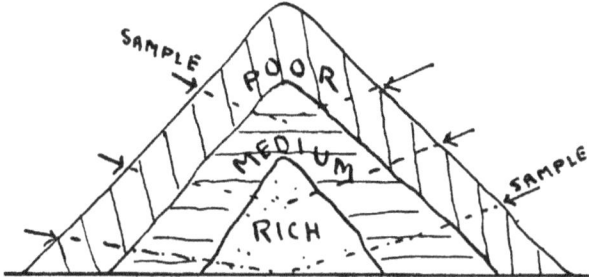

fines were dropped in one place and the crushed ore from the breaker in another. It was found that the two classes did not mix to any extent in the bins but were discharged, each by itself, into the automatic battery feeder. The feed was therefore very irregular, with the consequent breakage of stems. In all well-designed mills to-day the grizzly discharge meets the rock breaker discharge, or the grizzly is rejected altogether.

Another case: In running a silver leaching mill it became necessary one month to increase the output of silver. The higher grade ores in the mine were, therefore, drawn upon. The mill bin and crushed ore bins were about half full of twenty-two-ounce silver ore. The higher grade ore, containing about fifty ounces to the ton, was dumped on the top of the ore in the mill bin. For ten days after the richer ore was sent to the mill, the vat charges ranged from twenty to twenty-four ounces, but all at once they jumped to forty-five ounces, then to forty-eight, and then to from fifty to fifty-three ounces, showing that no appreciable mixing of the ores was made. It will then be obvious that mixing in a machine will simply mix the small amount of ore passing through it, and that the effect, so far as mixing the ore in bulk is concerned, is nil. Mixing is an ignis fatuus that we need not follow. There is, however, another class of engineers whose arguments may be respectfully considered. These say, granting that no mixing of the ore in bulk is possible, yet it must be obvious that if a stream of ore going through a sampler is thoroughly mixed, then any errors that may be inherent in the machine, will be counteracted. At first sight this seems a fair and sensible argument. It will not bear analysis.

Any machine that is so imperfect that requires an auxiliary to overcome its faults has no place in modern works, and should be consigned to the mechanical boneyard. Besides, a single machine properly designed and constructed will work more reliably than a "compound" machine.

CHAPTER VI.

Machine Sampling.—Machines for sampling ores, although of great number, may be divided into two classes:

1. Machines which take a portion of a stream of ore all the time.

2. Machines which take the whole of a stream of ore a portion of the time.

Since all sampling machines cut a stream of ore, it is of importance to consider the manner in which the cuts are made. It will be obvious that if a stream of ore is quite uniform in value a small portion taken out at any moment and from any part will give a perfect sample of the whole. This we know, not to be the case; the stream of ore changes in value at irregular intervals. It is, therefore, necessary to make the cut at frequent intervals, and in such a way as to get an accurate sample. Hence, the necessity of understanding the laws which govern the flow of a stream of ore is very clear.

Another matter to be considered is "probability." A great deal is made of this term, which is in fact out of place in transactions such as ore buying and selling. The miner certainly sells ore, and the buyer pays money, and there is no probability concerning these transactions. It is necessary, therefore, for each to certainly receive what is agreed upon. The idea underlying the term probability, is that, supposing the sample obtained from one lot is a dollar or two too low, then the next sample will be a dollar or two too high, and so the results will balance. So far as the buyer is concerned, this probability works out satisfactorily. If, however, a miner sends 100 tons of ore to the buyer, and he receives $200 too little for his ore, it is scant comfort for him to know that the next fellow will get $200 too much. Of course, if he is the next fellow, it is all right, but should he never send another lot of ore to the buyer, then it is obvious that he is a loser by $200. A machine which only "probably" gives a correct sample is worthless as a basis for money settlements.

Correct ideas as to the flow of ore in streams will greatly conduce to more accurate ideas as ta the relative values of samples and sampling.

The salient points concerning such streams are:

1. The ore stream in the direction of its flow changes

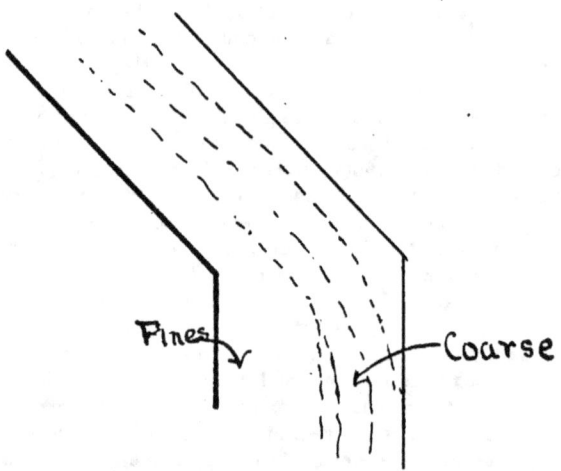

gradually from rich to poor and vice versa.

2. The distribution of values across the stream is exceedingly irregular, and the differences between the different parts of the stream very great.

A number of machines, especially some of those which take a portion of an ore stream the whole of the time, depend for their efficiency upon the uniformity of values across the stream. This point is so important that it will be considered first.

In a stream of ore flowing down a shoot the coarse pieces rise to the top, the fines travel along the bottom and against the sides. Now, whenever such a stream of ore changes in direction, the large and heavy pieces will tend to travel or shoot over to one side of the spout owing to their greater momentum. The stream of ore is then flowing in two parallel streams, coarse on one side and fines on the other. Such streams mix but slowly.

The demonstration of the fact that parallel streams do

not mix can be shown readily in every placer mining district where a stream, on which placer operations are under way, joins a clear stream. Both will flow side by side on the same bed, but quite separate for considerable distances.

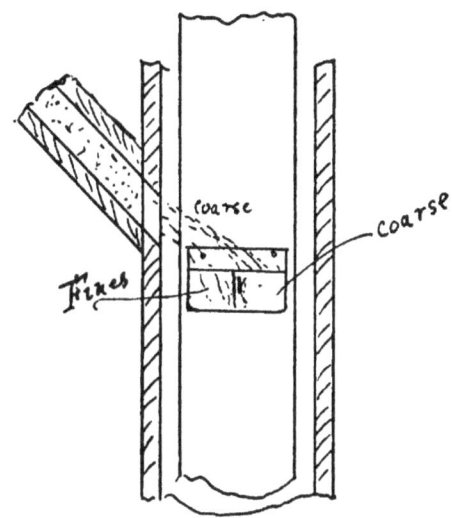

The confluence of the Saone and Rhone, in France, is quite a noteworthy case. The former, which is muddy, joins the limpid Rhone, and runs with a clear line of demarcation for a number of miles. A stream of ore acts in just the same way; if rich ore flows on one side of a shoot and poor ore runs on the other, then they will mix only slowly, if at all.

Besides changes in direction, many other causes may effect the same results:

1. Two elevators throwing ore into the same shoot; one elevator raising poor ore, the other richer ore. If the streams thrown from each do not cross, then little mixing takes place. If the streams cross, mixing is more complete.

2. Feeding an elevator from the side instead of from the front.

3. Irregular feed of coarse and fine to a rock breaker. Coarse on one side of jaw opening, fines on the other.

4. Anything which will produce an eddy will cause separation between the different sizes of ore in a stream.

It is well known that fines and coarse are, almost in-

variably, of different values, and hence whenever a stream of ore has more fines on one side than on the other, then one side of the stream will be relatively richer or poorer than the other.

The stream of ore in direction of its flow, changes, but, as a rule gradually, from rich to poor or vice versa. For example:

Take a stream of ore, which is, say ten feet in length. The probability is that the first and last foot of such a stream will differ in value, more or less. The stream will not change abruptly, but will merge from one grade to another. Thus, suppose an ore of a value of $10 per ton is being run through the mill, and that immediately following it is an ore of a value of $30 per ton. Although it is true that when these ores are run into the bin, there will be, for all practical purposes, no mixing, yet, in the shoot or spout itself, the ore stream will not abruptly change. The stream of $10 ore will gradually change, first to say $11, then to $12, and so on. Take every alternate foot, and with a stream of ore which is gradually changing in value, it will be obvious that S2 will be an average of 1 and 3; S4 an average of 3 and 5; S6 an average of 5 and 7, and so on. These samples mixed together will give an absolutely accurate sample of the stream for ten feet in length. So will S6 by itself, but the chances of cutting a sample at the exact place would be so small that it is necessary to make five cuts instead of one: For example, should cut No. 4 be taken, the sample would be too poor. Should No. 8 be taken, it would be too rich. The idea is that whenever a stream of ore changes in value from rich to poor, or vice versa, the change is invariably gradual.

It is obvious that a sampler which takes a longitudinal slice of the stream, can not possibly take a correct sample. It should also be pointed out that a slice taken across a stream of ore must be taken correctly if it is to correctly represent the two adjoining portions of the stream.

Suppose the sketch repesents a stream of ore, which is rich on one side and poor on the other. Then a sample taken as No. 1 will be far too rich, and if taken as No. 2 will be far too poor. The fundamental law of sampling has been violated, since more ore was taken from one place than from another.

By keeping strictly before us (1) the correct methods of making a cut from a stream of ore, (2) the characteristics of the flow of a stream of ore (3) the fundamental law of sampling, then we can have very clear and precise ideas of the efficacy of any sampling machine or process.

CHAPTER VII.

In Chapter VI the character of a stream of ore was fully considered. It was explained that since all mechanical samplers took portions of a stream of ore, that it was necessary to analyze the distribution of values across such a stream in order that one might be able to judge as to the correctness of a sample taken by an automatic appliance. Our engineer provides for us a very simple and satisfactory method of so doing. It may also be recalled that the consideration of the characteristics of the stream of ore showed conclusively that a good sample could not be made if only a portion of the stream were taken all the time.

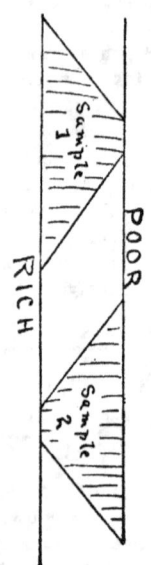

If we lay off on paper the lines taken by the course of the stream of ore, and lay off along those lines the time during which the scoop or other automatic appliance is taking the sample, and then across the stream, project the exact lines taken by the scoop as it crosses the stream of ore, we can then see exactly the shape of the slice taken. If this slice shows that as much has been taken on one side of the stream as on the other, then we may be sure that the sampler is at least working according to that first law of sampling laid down in these articles so many times. If we refer to the figure below, we see that although sample No. 1 and sample No. 2 are valueless, yet the unshaded portion between the two would result in a very good sample being taken, if the same were taken at frequent enough intervals.

It may be instructive to refer to a sampler which was believed for a long time to be a very perfect one. The sampler referred to is the old form of Brunton sampler. It will be remembered that in this form of sampler a deflector swung across the stream of ore, the edge of which traveled along the arc of a circle. It deflected the stream of ore, first on one side and then on the other. If this deflector started from the side of the stream marked in the cut as "poor" and traveled from left to right and then again back from right to left, a sample would be taken exactly as that marked "Sample No. 2," with the result that the

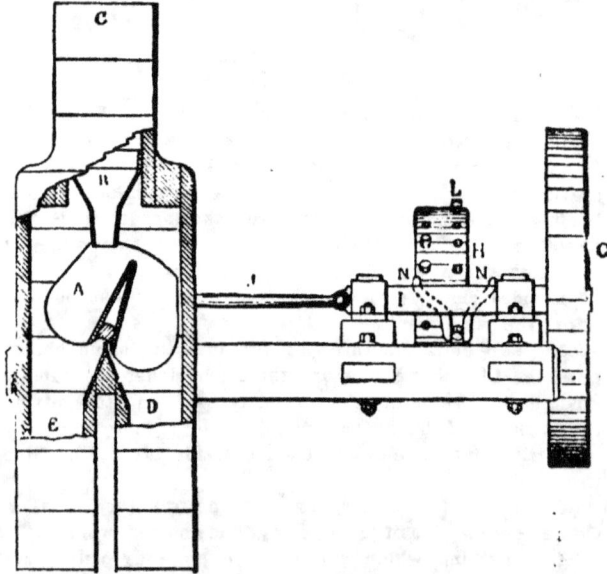

sample would be much too poor. It is, therefore, perfectly obvious that a sampler which takes its cut of ore alternately back and forth across a stream of ore in this manner cannot be accurate, and to-day there is no sampler that we know of of this character on the market. The most trustworthy sampler is one which always moves in the same direction across the stream of ore.

Although a sampler might give a perfect path of the

scoop across the stream of ore, yet in order for it to do its work perfectly the stream of ore should be kept as solid as possible. A falling stream of ore is difficult to control; it scatters and strikes the edge of the sampler with great force, and at all angles. But, by confining the ore so that it is delivered in a steady stream and at a minimum velocity, the correctly designed sampling machine will do its work quietly and according to the plans of the designer.

In addition to the principles already laid down as to the method of workng automatic samplers, in order for such machines to be really efficient, they should fulfil certain other requirements.

1. The machine should be simple, readily cleaned, stand much wear and tear, and not require to be stopped at frequent intervals for adjustment or repair.

2. As the quantity of the ore is reduced in bulk, the ore should be recrushed so that the ratio of the largest piece to the weight of the whole sample shall be within limits already laid down in a previous article.

3. The cut should be taken as frequently as possible.

And to recapitulate what was said in the beginning of this article:

4. The sample should be taken across the entire stream of ore.

5. It should be taken evenly from all parts of the stream, i. e., as much from one part as from the other.

6. That the stream should be delivered steadily to the sampler, and in as solid a condition as possible.

Any automatic sampling machine or plant built according to these principles will certainly give a most accurate sample in the quickest possible time, and at a very small expense.

CHAPTER VIII.

The Bridgman Sampler.—This machine was introduced to public notice about eleven years ago. Mr. H. L. Bridgman thus describes his machine:* This machine occupies a floor space of 3x4 feet and has a total height of 7 feet 6 inches. It is self-contained, requiring only to be bolted to the floor and to have feed, discharges and belt connections made. Fig. 1 shows the machine as it is built, while Figs. 2 and 3 give the diagraphic sections and details, some minor changes and omissions having been made for the sake of clearness. The machine consists essentially of three apportioners, 1, 11 and 111, all driven by the one pulley, X, and three stationary, concentric receptacles, R₁-R₂ and H, so constructed that any material falling into them will pass out through the spouts T₁ and T₂ into the sample buckets, Z₁ and Z₂, or through the spout S, which discharges the rejected portion of the sample. Apportioners 1 and 111 revolve in the same direction, apportioner 11 in the opposite direction; 1 at about 5, 11 at about 15, and 111 at about 45 revolutions a minute. That is to say, each apportioner moves actually three times as fast as the one above it, and in the contrary direction, or, relatively four times as fast. By the use of this expedient of contrary revolution, the same relative speeds are obtained as though, all revolving in the same direction, the actual speeds were respectively 5, 25 and 125, at which latter speed centrifugal force would become very troublesome.

The upper apportioner, 1, consists of two concentric rings, divided by eight partitions into eight equal topless and bottomless compartments L, from each one of which leads an adjustable spout, either as M-1 or as M-2, or as M-D. Set in rotation spout M-1 would describe a certain circular path, 1-1; spout M-2 a certain other path, 2-2, and spout M-D a third path, W (see Fig. 3).

*Transactions American Inst. Mining Engineers, vol. 20, p. 422.

The intermediate apportioner, 11, is merely a conical funnel, having, besides the large outlet W, four vertical shoots, N₁-N₁, and N₂-N₂, through its sloping sides, as shown in Fig. 3; each one of these shoots forms one-eighth of the circular paths covered by the spouts M-1 and M-2 respectively.

The lower apportioner, 111, is of the same construction as 11, and bears the same relation to it that 11 bears to 1.

An example will best illustrate the operation of the machine. It may be assumed that an original sample of 40,960 pounds (the 960 being added to avoid fractions) is to be put through the machine; that the time required will be one hour; that the speed of the machine is such that the upper apportioner, 1, will make 320 revolutions in that time and finally that the ore is of such grade and character as to only require the smallest sample that the machine will give. Under these conditions one of the spouts, M, would be set as M-1, one (the opposite one) as M-2, and the remaining six as M-D (Fig. 2).

The machine would then be set in motion and the flow of material, previously crushed to below one inch in size, would be started through the feed spout, F.

It is evident that at each revolution one 320th part of the whole lot, or 128 pounds, will pass through the feed spout, F. Of this amount six eighths, or ninety-six pounds,

A NEW SYSTEM OF ORE-SAMPLING.

Fig 1.

Mechanical Ore-Sampler. Machine A. General View.

will be discarded by the six spouts, M-D, passing down through W, W, H and so through the spout S and out of the machine, while one eighth of the 128 pounds, or sixteen pounds, forming the first cut of the first or outer sample will pass through the spout M-1, and the remaining one-eighth, or sixteen pounds forming the first cut of the second or inner sample, through the spout M-2.

These two first cuts will proceed side by side by separate paths through the same series of operations, and whatever applies to the one applies equally to the other; it will, therefore, suffice to follow the first sample. This one-eighth, or sixteen pounds, having been cut from the mass by the partitions of the compartment L, of which M-1 forms an extension, will drop nearly vertically through M-1 on its way to the sample box, Z_1. As it leaves the spout, M-1, during the one-eighth of a revolution that is occupied by the said M-1 in passing beneath the feed spout, F, it will be intercepted by the intermediate apportioner, 11, which in the same time will have made one-half revolution (relatively to 1).

Since the vertical shoot, N-1, occupies one-fourth of the semi-circumference of 11 passing beneath the spout, M-1, it follows that one-quarter of the sixteen pounds, or four pounds, will drop vertically through this shoot as the second cut of the first sample. The remaining three-quarters, or twelve pounds, will pass down the sloping sides of 11 and be discarded through W, W, H and S.

In precisely the same way the second cut of four pounds will be quartered by the lower apportioner, 111, three pounds being discarded and one pound, as the third and final cut, passing through the vertical shoot P_1, and the spout T_1, into the sample bucket Z_1.

In the same way a one-pound portion, as the third cut of the second sample, will find it way to the bucket Z_2.

This series of operations will occur at each revolution of the upper apportioner; and at the end of the hour each of the buckets Z_1 and Z_2 will contain 320 portions of one pound each, or a total final sample of 320 pounds, these two total samples being as independent of each other as though made at different times and places. It will, of course, rarely happen that this theoretical exactness of weights will obtain, which point will be considered later."

The Bridgman machine was taken up by a well known firm of mining machinery manufacturers, out in spite of all they could do to push it the machine has been declared to be unsatisfactory. It is a sad travesty upon the fallibility of engineers to quote language used concerning the machine at the time it was introduced. The words used were "it illustrates modern science and certainty while the methods it displaces may be compared with astrology in the vagueness of its conclusions. Investigate it."

"The first part of Bridgman's sampler takes a perfect sample; the rest of the machine is not of any value. He does not mix the 'eighths' and then quarter, then mix the quarters, and again quarter, but out of each eighth he takes a 'clip' either from a part of one end or from somewhere between the ends; and then from this alleged quarter he again 'clips' a similar piece or portion. He might just as well take his 1-128th at the very start, in fact, better, because it would always be 1-128tn, or very nearly so. His 1-128th of 40,000 pounds should weigh 312½ pounds.

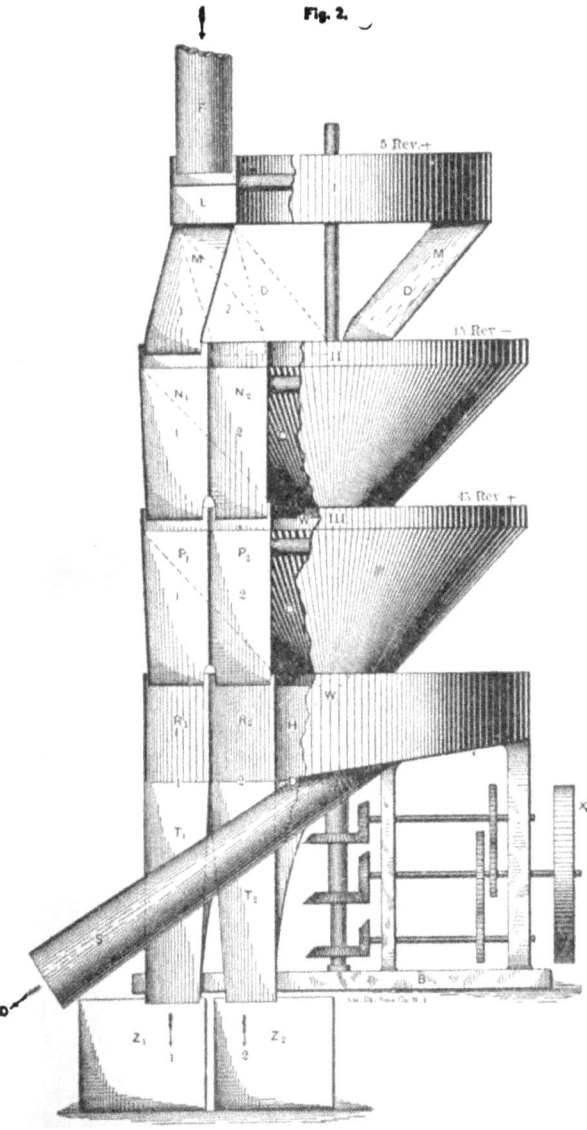

Fig. 2.

Mechanical Ore-Sampler. Size A. Total Height, including Sample-Buckets, 7 feet 6 inches.

By his table it varies from 101 pounds to 529 pounds. The sample is taken only once every twelve seconds, when all

good ones take a portion of the ore every one second to one every three seconds."

The fatal weaknesses in the Bridgman sampler are: (1) The lack of "mixing" between the apportioners; (2) the variability in weight of the samples obtained.

Lest any misconception should arise concerning the use of the word "mixing" used by the engineer, it should be explained that he really means a putting together of the samples. Our readers will remember that it has been shown: (1) That in a stream of ore the values fluctuate gradually along its length even if the ore is not homogeneous in value; (2) that mixing, properly so described, is an exceedingly difficult thing to do. Let us take a stream of ore flowing at the rate of eight feet a second, corresponding to a free fall of one foot only, and then take our first sample as in the Bridgman machine once every twelve seconds, then the slices are taken at intervals of ninety-six feet. The slices so obtained may vary enormously, both as to character and value. To sample an intermittent and variable stream of ore, such as slices passing through the machine at intervals of twelve seconds, is a very difficult thing to do. By throwing these slices, however, together and again drawing them out into a long stream, the conditions in the original stream—changing gradually from rich to poor—will again be obtained. By mixing our engineer really means what we have just indicated.

The variability in weight of the samples obtained shows that the machine does not work with certainty. The machine, if set to take a 1-128th part, should give, from 40,000 pounds of ore, 312½ pounds, and if working with anything like accuracy should have given duplicate samples of almost identical weights. In actual work some of the samples obtained had the following discordant weights, according to Mr. Bridgman himself:

	Outer	Inner
No.	Sample.	Sample.
1	170	413
2	526	194
3	473	191
4	190	327
5	300	315
6	471	211
7	238	360
8	462	447
9	101	166
10	500	529

Out of the whole twenty samples, only two were approximate to the calculated weight, and the difference between the weights of the samples was, in some cases, as in No. 2, more than 250 per cent.

This variation in weight naturally suggests the query if the weights are so variable how about the correctness of the samples?

Mr. Bridgman gave in the paper, already quoted, some figures relating to the working of the machine. "It was set below and takes its feed directly from a No. 2 Gates crusher, whose capacity, large as it is, is very much less than that of the sampler. The product of this crusher will not exceed three-fourths inch ring gauge in size. No attempt whatever was made to secure any special regularity of flow, the man at the crusher feeding as may be most convenient for himself. The discarded portion of the sample was removed by a bucket elevator to storage bins, the two final samples (of 200 to 500 pounds each) alone remaining for further handling. These final samples are often obtained within half an hour after the last of the material is out of the car. They should be crushed to one-quarter inch or one-eighth inch ring gauge and run through the smaller sampler. Having, however, heretofore had neither fine crusher nor small sampler, we merely quartered them down by hand to, say, twenty pounds, passed them through a laboratory crusher, cut them down again with sampling tin and "bucked" them; the ground material being finally mixed and distributed by means of the mixer and divider previously described.

Prior to the introduction of this machine the works had treated fifty-four carload lots (about 30,000 pounds each) of copper matte, on which double samples were made by hand. The average assay contents of these fifty-four lots were 7.88 ounces gold, 168.71 ounces silver, 55.24 per cent copper. The average differences between the two samples of each lot were 0.48 ounces gold, 3.77 ounces silver, 0.71 per cent. copper.

Since the introduction of the machine, twenty-two lots of ore and 138 lots of matte have been run, the latter being of the same general character as the hand sampled matte, except that it did not, as a rule, carry so much "metallics." By reason of these "metallics" much of this matte was very difficult to sample accurately, as will be easily understood. The weights of these 160 lots varied from sixty-five pounds to 42,000 pounds, averaging not less than 30,000 pounds.

Their average assay contents were 0.71 ounces gold, 112.04 ounces silver, 51.75 per cent. copper.

The average differences between the two samples of each lot were 0.02 ounces gold, 1.19 ounces silver, 0.23 per cent copper.

Reduced to percentages for the sake of comparison we find the average differences as follows:

	Gold Ounces	Silver Ounces	Copper per cent.
Fifty-four hand samples............	5.46	2.24	1.29
One hundred and sixty machine samples	2.82	1.06	0.45

The writer of the paper, H. L. Bridgman, called this a creditable showing, and at first sight it would appear so.

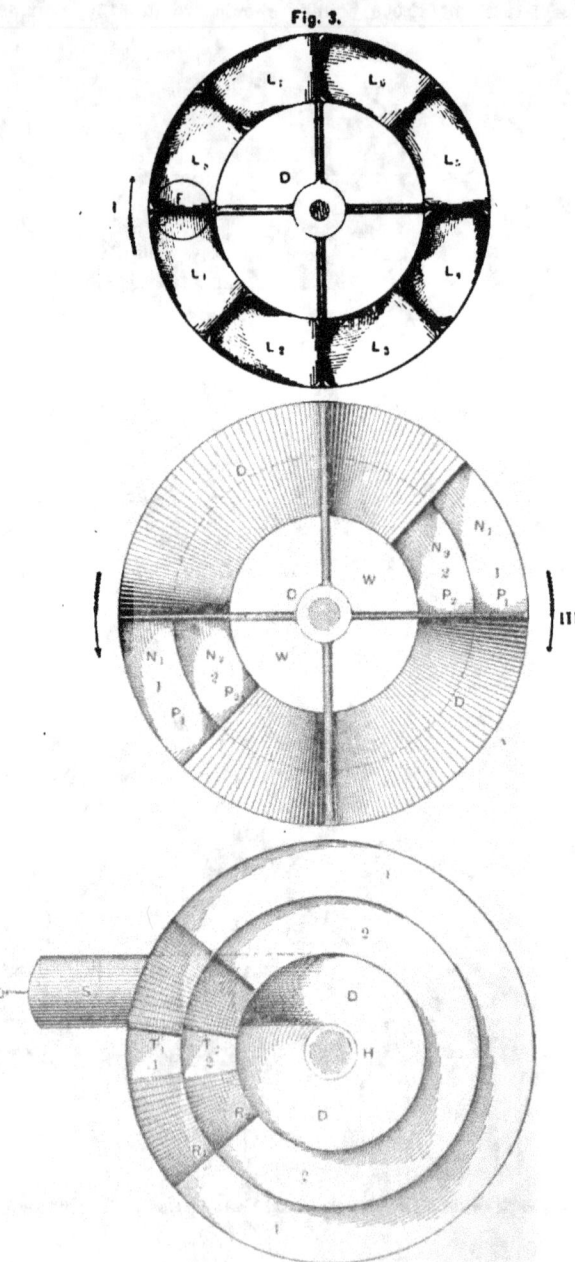

Mechanical Ore-Sampler. Size A.

The difference, however, between the values of the different lots of matte, the comparison between an archaic method and a new machine, the uniformity of the material sampled and the lack of complete information all destroy the value of the figures given. Our readers will remember that the subject of averages was discussed, and it was shown that, however complacently the buyer might regard a close average, yet the individual seller requires individual accuracy of sampling. It does not matter to him how the average might work out, and it is no comfort to him to know that if he is getting $500 too little for his ores that the other ore seller is getting $500 too much. The average results obtained are of little value in estimating the value of a machine designed to sample ores for the public, and it would be more to the point if the actual differences between the samples had been given.

It has been pointed out that comparisons to be of any value must be between similar things. To record a close agreement between a notoriously inaccurate method and a supposedly accurate one is to condemn the latter. The comparison between the coning and quartering method and the Bridgman machine is of no value.

Another feature which destroys the value of the figures given is the uniformity in character of a given consignment of copper matte, and there is no difference in value between the fines and the coarse in precious metals or copper in crushed matte. Figures obtained from sampling such material are not worthy of attention, when the question under consideration is the sampling of such ores as telluride gold ores, or rich chloride and sulphide of silver ores.

The Bridgman sampler has proved to be unsatisfactory in sampling ores, although from a cursory examination it might appear to be a most excellent machine. The analysis of a stream of ore passing through the machine discloses instantly why it gave such unsatisfactory results.

CHAPTER IX.

The analysis of a stream of ore passing through the Bridgman sampler is of the greatest interest, and the marked agreement between the results as shown by the facts disclosed by the analysis and the actual results obtained in practice, show that the method of analysis adopted is correct.

In the Bridgman sampler the first cut is made with a fair degree of correctness, since the manner of taking the cut fulfills the laws already laid down, viz:

(1.) The cut is made across the entire stream of ore.

(2.) The sample, or cut, is taken evenly from all parts of the stream.

(3.) The stream of ore can be delivered steadily to the cutting device, and in a fairly solid condition.

It fails in one particular only:

(4.) The cut is not taken frequently enough, being taken at intervals of about twelve seconds, whereas all modern samplers take the cuts at intervals of from two to three seconds.

The work of the second apportioner, however, fails to do good work, and the third apportioner does even poorer, for

(1.) The ore is not delivered in a steady stream, but goes through in sections, and, therefore,

(2.) The cut may or may not be made evenly from all parts of the stream; and,

(3.) The cut may only represent a sample of a part of the stream.

The cut, however, is made at more frequent intervals than that made by the first apportioner.

In order to analyze the work of the machine it is necessary to calculate:

(1.) The number of cuts made per minute by each apportioner, and the time of taking the cut; and,

(2.) The velocity of the stream as it reaches each apportioner; and,

(3.) The condition of the stream of ore passing each apportioner, and then, finally, to calculate,

(4.) The shape of the cuts made by each apportioner.

These facts may be estimated from the following approximate details of the machine:

At the first apportioner the ore has fallen eighteen inches and has consequently a velocity of 9.8 feet per second. The apportioner has a mean circumference of two feet, and makes five revolutions, and, therefore, has a velocity of thirty-one feet per minute. A cut is made every twelve seconds and is taken for one and one-half seconds.

At the second apportioner the ore will have fallen three feet, and will have a velocity of fourteen feet per second. The apportioner has a mean circumference of two feet six inches, and makes fifteen revolutions in an opposite direction to No. 1. The mean velocity of the openings in the apportioner is 107 feet per minute. A cut is made every two seconds, and is taken for one-half a second.

At the third apportioner the ore will have fallen four and one-half feet, and will have a velocity of seventeen feet per second. This apportioner has the same circumference as No. 2, but makes forty-five revolutions per minute in an opposite direction to No. 2. The mean velocity of the openings is 351 feet per minute. The cut is made every 0.65 of a second and the cut is taken during 0.16 of a second.

We will now calculate the conditions under which each cut is made. We will asume that a steady stream of ore is being sent to the first apportioner. Every twelve seconds a piece of the stream will be cut off which will be 14.7 feet from one extremity to the other. Ten and one-half seconds after the tail end of the first cut has passed through the scoop another cut will be started on. The diagram below shows the manner of taking the cuts, the shape of each cut, and the condition in which they are sent to the second apportioner.

We thus see that the conditions which exist at the second apportioner are radically different from those at the first, since:

(1.) The ore is not delivered in a steady stream, but in pieces, or sections, of a stream at intervals.

(2.) The intervals between the sections are quite wide apart, viz, 10½ seconds.

(3.) The character of such an intermittent stream is very different to that of a steady stream.

The last point we will consider later, remarking, however, that it is of extreme importance. We will follow first the course of the ore through the sampler.

For the second apportioner to take a sample at all, it depends upon whether the opening of that apportioner is passing under the feed spout whilst ore is passing through it. The section of ore passes every 10½ seconds and the stream lasts from one and one-half to one and three-quarter seconds. Between each cut of the upper apportioner, therefore, each scoop of the second apportioner passes five and one-quarter times under the feed spout. The chances are, therefore, that ore may pass through the spout and none of it get into either of the openings. On the other hand, they may be in such positions that each may take a portion. Hence we may have these conditions:

(1). One-eighth of the ore may be taken, or less, according as one of the openings may completely pass through the intermittent stream of ore.

(2.) A cut may be taken varying from one-eighth to one-quarter. The amount may vary between these p oportions. The shape of the cut taken will depend upon whether the opening is early, late or exactly on time in making the cut. As shown in the diagrams the cut may assume any one of three shapes, modifications of those shown.

It is obvious that, since the stream is passing during only from one and one-half to one and three-quarter seconds, and the opening is only passing through the stream for one-half second, that a "clip" only is taken from the stream, and it depends very much as to the point of the stream taken, whether a fair sample is taken or not.

Now we come to the second apportioner. About every twelve seconds a stream of ore will be delivered to it which will last about one-half of a second and which may be followed two seconds later with another stream lasting from a mere fraction to one-half of a second in duration. The stream which was sent to the second apportioner, irregular as it is, is as regular as clockwork compared with that sent to the third. Each opening, in the third apportioner passes eight times under the feed spout between the feeds. The time the ore is passing is only about from one-half to three-quarter seconds. At intervals of 0.65 seconds, one of the openings of the third apportioner comes under the feed spout, and as the stream only lasts for about one-half of a second, the scoop may not be there to get any of the ore. The shape of the cut will resemble somewhat those made by No. 2.

Thus we see that the sampler works with great lack of uniformity, and there is every chance of it either taking too much or too little.

The reason is now obvious why the weights of the samples disagree so much in Mr. Bridgman's samplings, varying, as we have seen in some cases as much as from 526 pounds to 194 pounds in the duplicates. The calculated weight was 312½ pounds.

We have seen why the weights of the samples disagree on duplicates and it now remains to see why the samples obtained are, at least, open to suspicion.

In a previous article the character of a stream of ore was fully considered. The considerations, however, were confined to a steady stream of ore, and although the facts brought out in that article apply also to an intermittent stream, yet there are many points in an intermittent stream that do not apply in a steadily flowing stream.

If a stream of ore is allowed to fall freely for a short space of time, we find that the coarse lumps will fall much faster than the smaller pieces or the dust. Then this condition of affairs exists: The ore first reaching a given point will consist mainly of large pieces, whilst the last part of the stream will consist of fines. The shorter the time of flow, the more marked will be the differences in the character of the first half of the flow and the latter half.

The ore stream leaving the first apportioner flows for one and one-half seconds, and a stream equal to 14½ feet in length will be delivered to the second apportioner. The ore stream leaving the second apportioner will be about seven feet in length, and will flow for about one-half of a

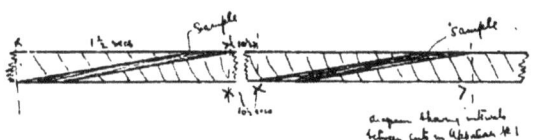

second to the third apportioner. In both cases the first foot of the stream will consist of an undue amount of coarse —in fact will consist almost entirely of coarse—the last foot of the stream almost entirely of fines.

Now comes the question, how does the sample opening take the cut from the first, middle or last portion of the stream? It is practically impossible to answer this question, but it would seem that it would depend upon the relative positions of the openings when the ore first commences to flow. It may take more ore from the top at one time, more from the bottom at another; at others, more from the middle. The machine works hap-hazard. In the long run, the results might average up for the buyer, but

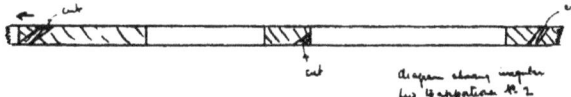

the correctness of the sample obtained from any one particular lot of ore is very uncertain, indeed. As pointed out time and time again any sampler designed to handle public ore must give on each lot a correct sample, and there should be no possible misgivings on the part of the ore seller as to the correctness of the sample obtained. Finally, as pointed out in the previous article, the sampler fails because the slices of ore are not thrown on each other and again drawn out so as to get the gradual change from rich to poor as it does in a steadily flowing stream of ore.

CHAPTER X.

Pipe Sampler.—One of the earliest forms of automatic samplers that came into use was the pipe sampler. The construction of this appliance is shown in Figure 1. In using this that came into use was the pipe sampler. The construction of this appliance is shown in Figure 1. In using this appliance the crushed ore was fed into the hopper a, which delivered the material through the narrow funnel b, to a split or divider at c. The stream of ore was divided by the split into four equal parts as in the case of coning and quartering. The opposite quarters are retained in the pipe whilst the other two quarters are delivered to the ore bin d. The two quarters remaining in the pipe are concentrated in the funnel b; and again sent to another split or divider. The opposite quarters are again rejected, and this operation is repeated as often as may be deemed necessary in order to obtain a sample of sufficiently small size, which went to the sample box e. The rejected portion of the ore was caught in the bin d, from which it could be drawn through the gate. The details of the pipe sampler are shown more fully in Figure No 2.

This sampler was for a long time a favorite one. The arguments advanced being. first, that no mechanism is required, and secondly that because it takes a sample continuously, a perfect sample must be obtained. The device, however, has so many disadvantages that it is no longer used where accurate sampling is required. It has long been recognized that the employment of power in itself is of no particular disadvantage, and if by using power a more uniform character of work is obtained, it is advisable to use it. The old idea that a perfect sampler is one in which a sample is taken all the time, we have already disposed of. It may, however, be pointed out that the idea of the machine taking a sample continuously is a delusion, and if a reference be made to article No. 6* it will be readily seen that taking a vertical cut from a stream of ore, gives at best a very uncertain and unsatisfactory sample.

.n analyzing the work of the sampler, it is merely necessary to consider the work of the first split, because if the first division is not correct, then no degree of exactitude in the latter operation will insure a correct sample. The difficulty in such a sampler as the one we are now discussing is to present a separation between the fines and the coarse. If the coarse is fed into one side of the hopper, and fines on the opposite side, then the splits will take out an undue proportion of one of the other. If the coarse and fines have dissimilar values, then an incorrect sample will

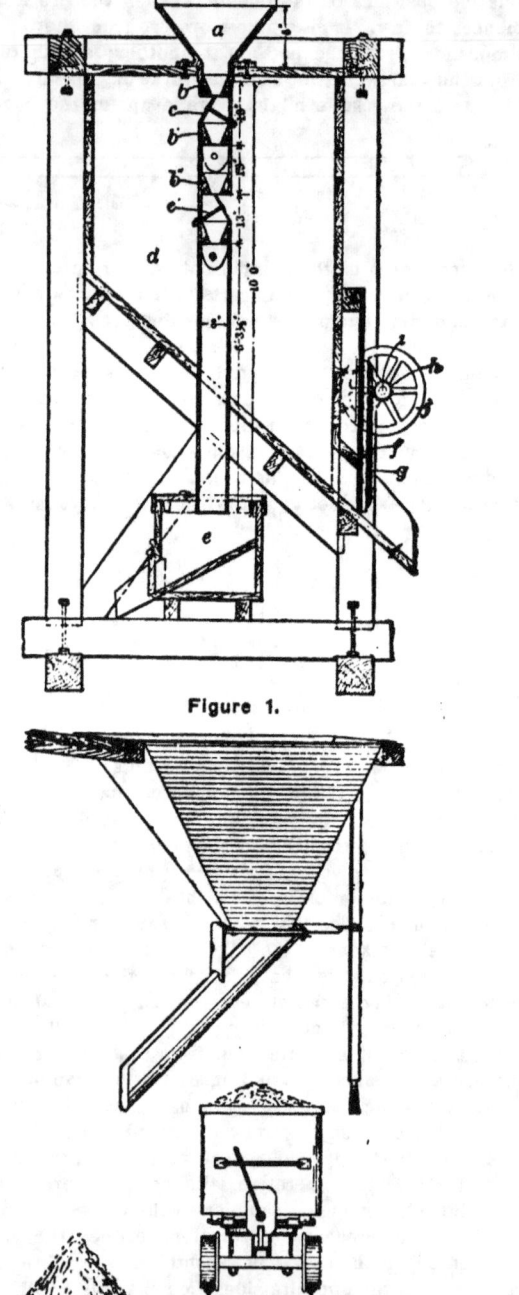

Figure 1.

Figure 3.

be obtained, and if this sample be further cut down, this inaccuracy cannot possibly be rectified. Of course, in some furnace products, like matte, the difference in value between the fine and the coarse is almost nil, and in such cases as these, the pipe sampler might give a fairly satisfactory result. But it might be asked, why should an uncertain appliance be used when there are perfectly satisfactory samplers now on the market, even if only furnace

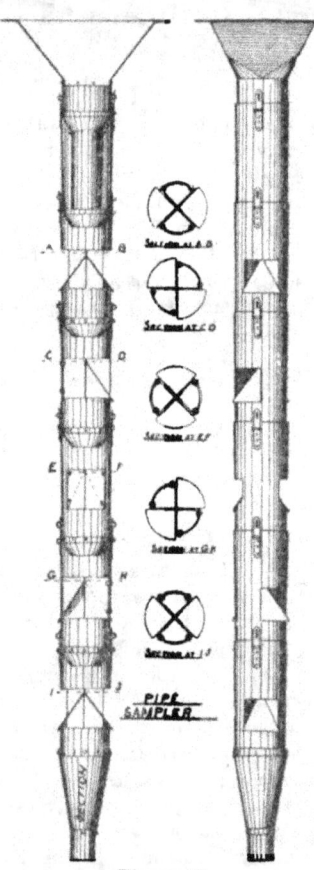

Figure 2.

products are being sampled? For sampling ores, for purchase, the pipe sampler is open to grave objections.

Single Split Sampler.—This form of sampler is even yet more inaccurate and unreliable than the pipe sampler. It is illustrated below in the form as it was formerly used. It will be seen that this "sampler" consisted of a small storage hopper at the bottom of which was a splitting arrangement. Whenever the gate of the hopper was opened to fill a car, or wheelbarrow, a splitting arrangement cut out a portion of the stream of ore, and delivered it to one side. This sampler is simply mentioned for the purpose of condemnation. It has no merit to recommend it. In fact, we might say that large smelting works which purchased Cripple Creek ores for over a year, through using a similar device, the split sampler described in our issue of December 12th, lost, it is said, upwards of $300,000. We mention this because it shows how a comparatively small cause of error can make such tremendous losses in the buying and selling of ores.

*Mining Reporter, December 26, 1901.

CHAPTER XI.

The Brunton Sampler.

Mr. D. W. Brunton has invented two sampling machines, the earlier of which gave, under certain conditions, very inaccurate results. The newer Brunton sampler works on an entirely different principle to the old machine, and gives a very satisfactory sample. It is very desirable to examine into the merits of any satisfactory machine, but it is of equal, if not greater, importance to learn why a machine has failed to do reliable work. With this in view the old Brunton machine will be first described.

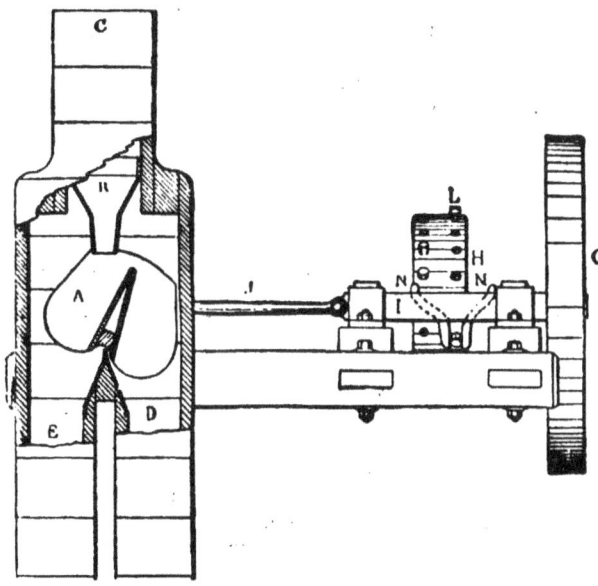

Brunton's Old Style Sampler.

The original Brunton machine worked by deflecting a falling stream of ore to the right and left first into one spout and then into another. For example, if a sample was required as large as one-half of the total amount of ore the stream would be deflected one-half the time to the sample spout. If only twenty per cent. of the total were required then the stream would be deflected into the sample spout a time which would amount to one-fifth of the time the ore was falling. The cut was to be taken usually every three seconds. If the stream were falling during a period of one hour then 1,200 cuts would be taken. When the machine was first introduced it was highly praised, but it is to be feared that the investigators were very superficial in their examination. After long continued trials the machine was condemned and the inventor himself is one of the severest critics. He was so dissatisfied in fact that he discarded the machine and designed a new one which works very satisfactorily.

The old Brunton machine is deficient in two respects:

(1.) That it takes more ore from one side of the falling stream than from the other.

(2.) That if it becomes necessary to take a larger sample it does not take the same quantity more frequently, but simply takes a larger quantity at the same intervals, that is every three or more seconds.

These objections are fatal to the usefulness of this machine. Before, however, demonstrating these points it may be as well if the machine itself were first described. In the illustration C is a vertical or inclined spout containing a falling stream of ore; B is a funnel which narrows up the stream so as to reduce the length of the necessary arc of travel of the deflecting chute A. The deflecting chute A is rocked right and left by a crank (not shown in the illustration), which is actuated by a reciprocating driving bar J. J obtains its motion through a drive bar I, which is reciprocated by cams or pins, L, in the driving wheel H. These pins act against curved pieces of metal, N, which act as tappets. The two rows of holes in the driving wheel are bored just such a distance apart as the drive bar is required to travel. There are twenty holes in each row, so that each hole or pin represents five per cent. of the time required for the complete revolution of the wheel. If a sample is required as large as fifty per cent. of the ore half the pins would be left out on one side of the wheel and half on the other. Then the revolution of the wheel will hold the chute, A, on the right during half the revolution and on the left during the other half. If twenty per cent. of the pins are placed in the right hand row and eighty per cent. in the left then the deflecting chute will be held on the right during one-fifth of a revolution and on the left four-fifths of a revolution, thus throwing twenty per cent. of the ore into E and eighty per cent. into D.

The method of analyzing the action of this sampler has been already indicated in a previous article. It may be as well, however, if the analysis of this machine were performed in detail. The analysis is made of the working of a machine built by Hendey & Meyer and which took its cuts from a four-inch stream of ore; taking about fifteen per cent. of the ore for a sample. The velocity of the moving ore is taken as if it had a free fall of one foot, or about eight feet a second. The throw of the driver, or deflecting chute, is seven inches. A cut is made every three seconds.

The travel of the deflector relatively to the falling stream of ore is shown in the accompanying figure as A, B, C, D. The deflector passes from A to B in 0.24 seconds.* The deflector remains at rest for 0.06 seconds and then moves back through the stream, the edge describing the path C, D. The shape of the slice of ore taken is shown by the figures A¹, B¹, C¹, D¹. It must be obvious that if one side of the stream of ore were richer than the other then the sample would be either too rich or too poor.

It might be denied that such a condition, as one side of the ore being richer than the other, could be permanent and that it is more reasonable to suppose that the values at the side of the stream would fluctuate, i. e., be rich at one time and poor at another. This matter has been gone into with considerable detail in a previous article,* but it may be recalled that such a condition as above indicated is a result of unskilful design, and is, therefore, likely to be permanent and to irregularities in feeding to the first crusher by a workman.

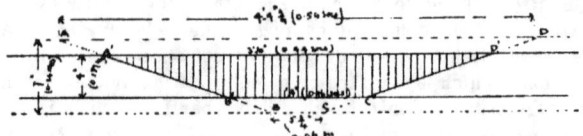

It may be stated that the frequency with which a cut

*Other figures are given in the diagram.
*Mining Reporter, December 26, 1901.

is made has no corrective tendencies if one side of the stream is always richer than the other. If the values fluctuate from side to side, then a frequent cut increases the chance of getting an accurate sample. We know that a flowing stream of ore changes gradually from rich to poor and vice versa in the direction of the flow and therefore that by taking cuts frequently a sample will be taken accurately if as much is taken from one side as from the other. But when another uncertainty is introduced the sample is doubtful, or at any rate both buyer and seller are much more dubious as to the correctness of the sample obtained.

It must be admitted that by careful feeding of ore to the first crusher, skilful arrangement of the plant and attention in running it that such a machine as we have just described will give fairly satisfactory results. But there must always be an element of doubt as to the work of a sampler which takes more ore from one side of a stream than from the other, and further, a machine which requires so many conditions becomes exceedingly unsafe.

The correct shape of a cut is shown in Fig. 3, between the two unshaded parts, where as much ore is taken from one side as from the other. Many machines are on the market which make a cut of the shape shown and these are preferable, all other things being equal, to a machine which makes a cut like the old Brunton sampler; this latter machine also violates the first law of sampling, already repeated many times, inasmuch as it takes more ore from one part of the stream of ore than from the other.

Addendum.—In describing his old machine Mr. Brunton, in the patent specification of his new machine, says: "A third plan has employed a vibrating spout by which the

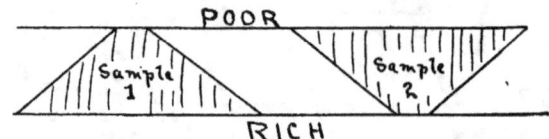

entire stream of ore is deflected alternately to the right and to the left. In this class of apparatus the dividing edge of the spout does not cross the falling stream instantaneously, but consumes an appreciable portion of time during which the ore has descended a considerable distance into the spout, thus rendering the sample section taken out of the falling stream a rhomboid or frustrum of a wedge, the base of which is taken from the side of the stream toward which the rejected ore is being deflected, and because of this operation it follows that unless both sides of the stream of ore are exactly alike the sample taken will not be an accurate or average representation of the lot."

The New Brunton Sampler.

In 1895 Mr. David Brunton designed and patented a machine which overcame the faults inherent in his first design. The new machine has worked successfully on the Cripple Creek ores, which are possibly the most difficult of any ores to sample correctly.

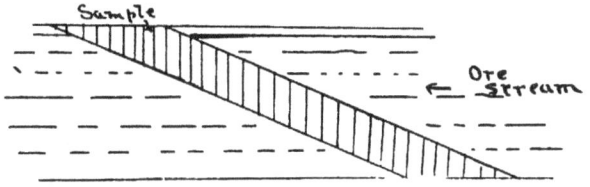

The sampler consists of a box in which works a double swinging spout, or scoop, which is so arranged that it cuts across the path of a falling stream of ore, giving a sample cut of the shape

In the old machine a single cutting edge mov-

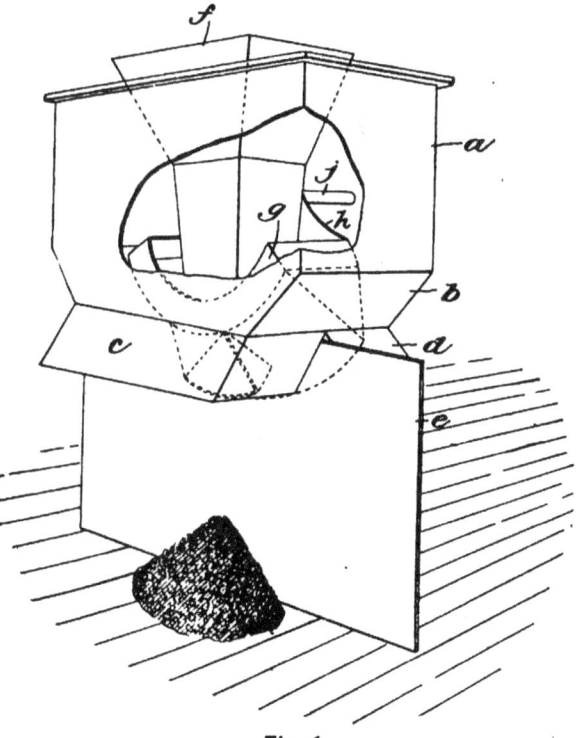

Fig. 1.

ing back and forth gave a more or less wedge-shaped sample cut; in the new machine the two edges of the scoop move across the stream and takes a clean, positive cut of

the shape above shown. The result, in a well designed plant in which a continuous and solid stream of ore is sent to the sampler, cannot fail to be satisfactory.

The following cuts, reproduced from the patent specification drawn up by Mr. Brunton, show the design of the machine. Fig. 1 is a prospective view, parts of the casing of the machine being omitted. Fig. 2 is a vertical transverse section. Fig. 3 is a vertical longitudinal section. Fig.

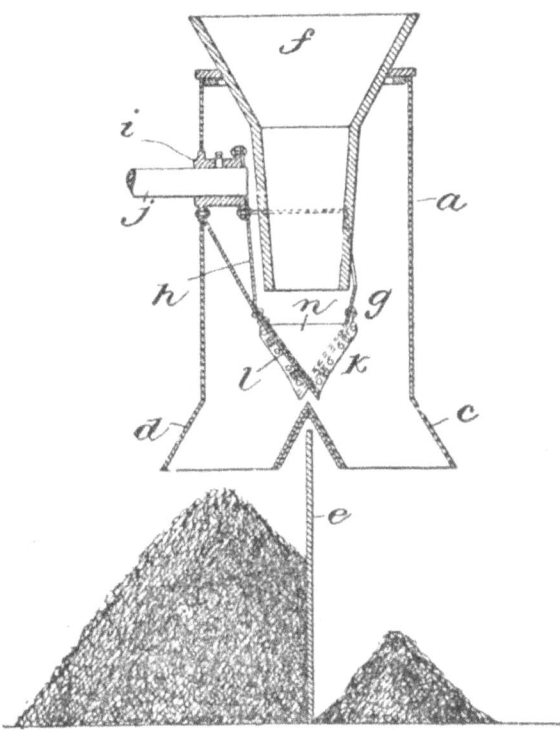

Fig. 2.

4 shows the details of the swinging spout, or scoop. The working of the machine is very simple; a steady stream of ore is delivered to the shoot f, which carries the ore to the deflector, g, which rocks to and fro across the ore stream. The deflector (see Fig. 4) consists of a back plate, h, and a pocket, or scoop, l, which is inclined in an opposite direction to the back plate. The deflector is, therefore, so arranged that the bulk of the ore is delivered to one chute, and forms the rejected portion of the ore, whilst when the pocket passes under the stream it cuts out a portion which is sent to another chute; this latter portion forms the sample. The details of the working of the machine can be readily seen by an examination of Figs. 2 and 3.

Mr. Brunton, in his specification, gives diagrams showing the shape of the cut taken by his new machine, as well as the shape of the cut taken by the old machine. These diagrams are reproduced in Fig. 5. It will be noticed that the cut taken by the new sampler is mathematically correct, being a rhombic prism. The diagram, which is said to show the cut made by the old sampler, is not correctly drawn. We show the actual shape of the sample taken by that machine in diagram A (Fig. 5) at a. It is a frustrum of a double wedge and is an incorrect shape for reasons pointed out in previous articles.

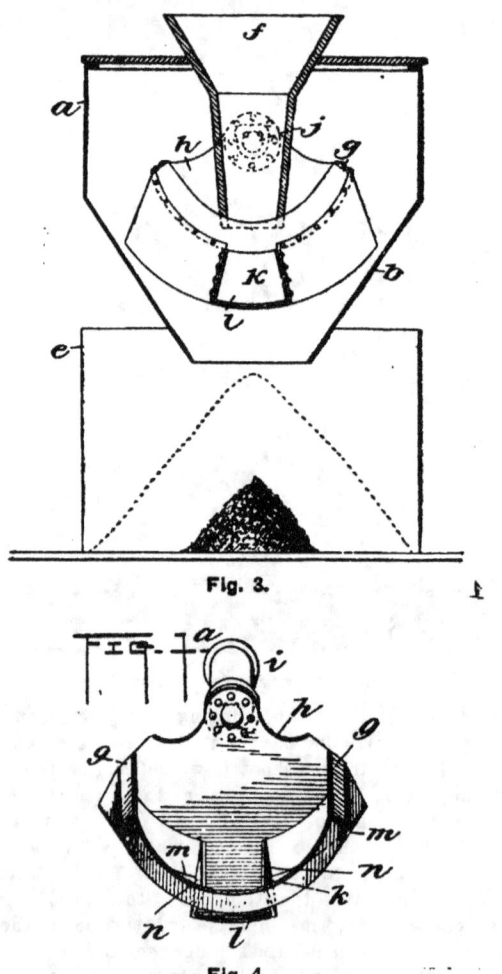

Fig. 3.

Fig. 4.

It is proper to say that the Brunton new sampler is one of the few sampling machines on the market which takes a correct sample. The machine, however, must be intelli-

gently used, and requires the sampling plant to be correctly designed in order for it to do good work. In the following figures we show sections through the Taylor & Brunton sampling works at Victor, Colorado.

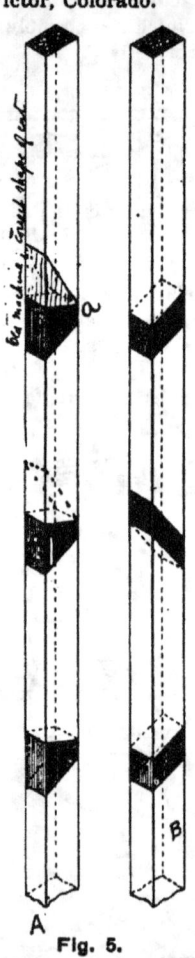

Fig. 5.

In the works referred to the ore is first fed to 15x9 Blake crusher, set to crush to one and one-half or two inches; the ore is delivered to an elevator, which runs at about 300 feet per minute. The elevator delivers the ore to the first sampler, S_1. The rejected ore goes to the ore bins, the sample proceeding to a set of rolls 36x14 inches, and is cut down finer by sampler, S_2; the sample is again crushed by a set of rolls, 27x14 inches, set to crush to one-quarter inch mesh. The one-quarter-inch material is again sampled in S_3. This sample is mixed and dried thoroughly and is sent to a fine set of rolls, 20x10, which crushes the ore to one-sixteenth inch, and sampled in S_4. Each sample takes out one-fifth of the ore as a sample so that the final weight delivered to the finishing department is one six hundred and twenty-fifth of the whole. This arrangement is very good and allows the sampling works to deliver to the smelters eighty per cent. of the ore received in the condition it leaves the rock breaker.

The capacity of these works is 250 tons per shift of ten hours. The entire plant cost, ready to run, $40,000. The power required is very small, being furnished by a thirty-five H. P. engine.

SECTION A-B

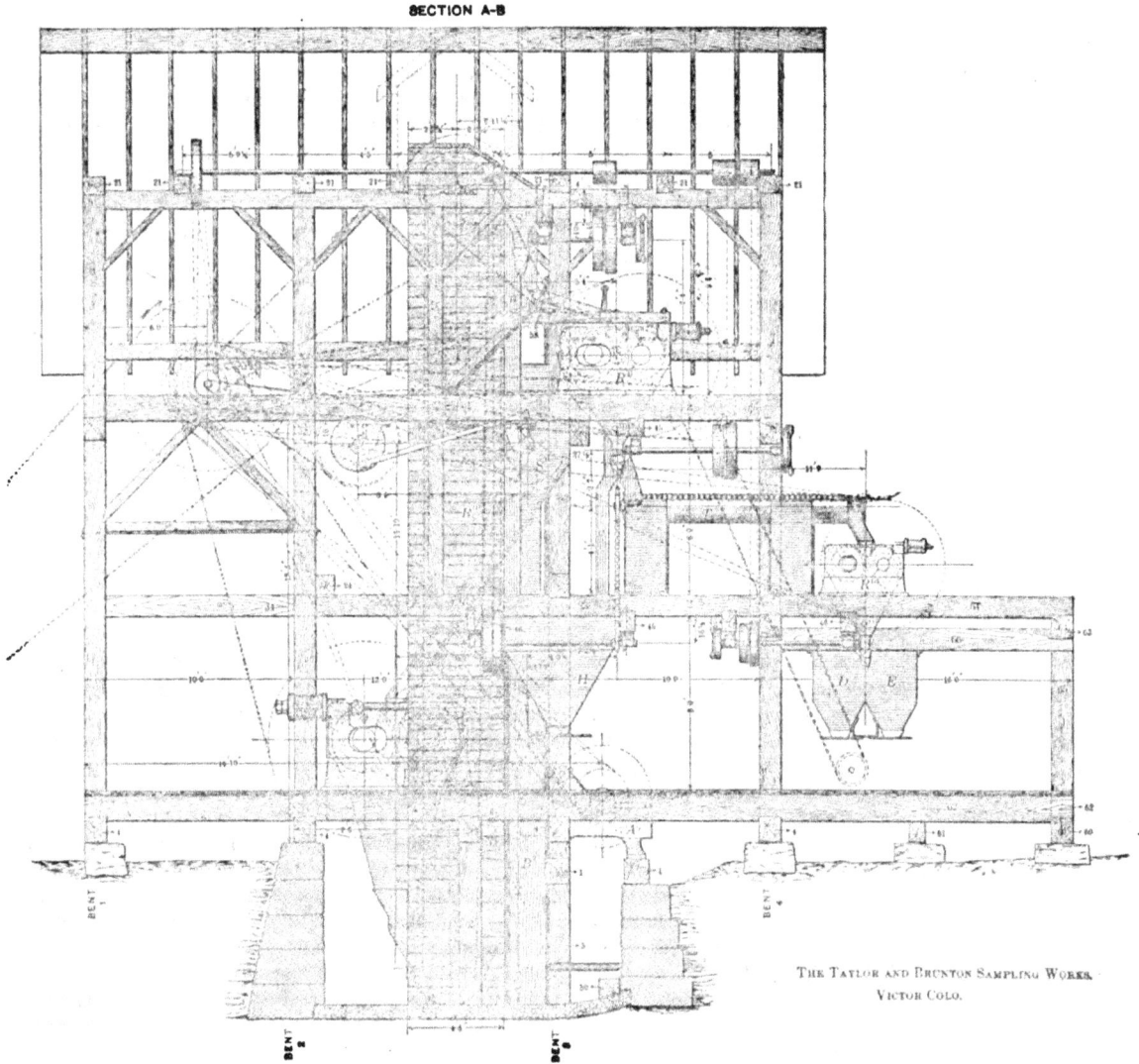

THE TAYLOR AND BRUNTON SAMPLING WORKS.
VICTOR COLO.

CHAPTER XII.

The principles underlying the method of taking a cut from a falling stream of ore in order to obtain a correct sample, are apparently not yet generally understood, since within the last year a machine has been patented which works on the identical principle used in the old Brunton sampler. The machine referred to is that designed and patented by Gustave A. Overstrom of Anaconda, Montana. By an ingenious arrangement in this design a deflector is moved across a stream of ore and held over for such a time as may be desired. The action of the machine is in no respect different to the old machine designed by Mr. Brunton. The work of this latter sampler was analyzed in our last issue and therefore it need not be again demonstrated that a deflector, moving back and forth through a stream of ore, gives an irregular cut. The Overstrom machine offers no advantage over the old Brunton sampler, and it is besides more complicated.

In the Overstrom patent specification a novel, but quite inefficient, modification of the "deflector" sampler is proposed. This variation consists in having a spout which delivers ore first into one chute and then into another. Whether the stream moves across a stationary deflecting edge, or the deflector moves across a stream of ore makes no essential difference in the general shape of the cut taken. In the case, however, of a stream of ore passing across a stationary dividing edge an irregularity enters into the analysis which does not occur in the other method. The irregularity is in the velocity of the stream of ore passing across the deflector. The cut is therefore a figure of the same general shape as in the other case, but instead of the cut being a frustrum of a wedge it is a more complex figure. At any rate more ore is taken from one side of the stream than from the other, and hence neither of the Overstrom designs will work well if there is any difference in the values between the coarse and the fines, or if the different sides of a stream have different values.

CHAPTER XIII.

The Vezin Sampler.—This machine was designed about thirty years ago, but it has only come into use during the last seven years. The Vezin sampler was recognized at an early date as being a thoroughly practical machine, and one which gave very accurate results. It shares with the new style Brunton sampler the distinction of being the standard sampler, and by many it is preferred to the Brunton. The Vezin sampler is not patented. The machine has been so largely introduced and has been used so steadily with satisfactory results during the last seven years that many more figures are available for this machine than any other and, therefore, an extended review of its work is possible.

The illustration shows the machine, set to take duplicate samples, in plan and elevation. The samplers, A and B, consist of two hollow truncated cones joined at their bases. The upper cone of each has attached to it one or more scoops, G, which, on the revolution of the double cone on its axis, C and D, take cuts from the stream of ore. The scoops are in the form of a sector of a circle. The sample cut is taken into the interior of the cone from whence it is conducted by a sheet metal pipe. The rejected ore shoots into a hopper and thence, by a pipe, is conducted to an ore bin or car. The upper cone, which carries the scoops, serves to prevent any stray pieces of ore from falling into the lower cone and thereby vitiating the sample.

The sampler is made of sheet metal, usually of 10 B. W. G. sheet steel. The wear and tear on the machine is surprisingly small. The size of the machine is governed by two requirements. 1. The size of sample required; and, 2. The size of the ore fed to the machine. It will be obvious that the size of the sample taken by the machine is fixed by the relative areas of the scoop and the circle described by the scoop during its revolution. If the area of the scoop is one-fifth the area of the described circle then the machine must take one-fifth of the falling stream of ore, presuming, of course, that the stream is regular, whatever may be the speed at which the machine runs. By running the machine at twice the speed one simply doubles the number of cuts, but the weight of ore taken at each cut weighs just about one-half, therefore the total weight of the sample taken must remain the same.

The size of the ore fed to the sampler governs the size of the latter by the following consideration: The minimum width of the scoop must be great enough to allow the largest piece of ore in the falling stream to pass through the opening with absolute freedom. The rule adopted by the designer of this sampler, in order that this condition may be met, is: "The width of the scoop at the center of the stream of ore should be at least four times as great

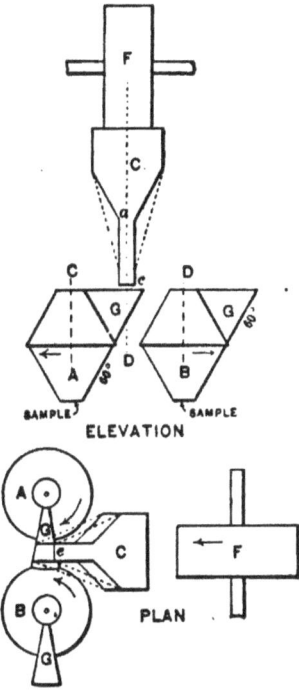

ELEVATION

PLAN

as the diameter of the largest pieces of the ore, and the width of the opening through which the scoop enters the upper cone should be at least 2½ times this same diameter to prevent choking—better three or four times." If the following conditions be given: Ore crushed to three-inch ring and amount of sample to be taken one-eighth of the whole then the width of the scoop at the middle must be twelve inches. The circumference of the described circle will be eight feet, and therefore the diameter of the machine from the axis to the middle of the scoop must be about thirty inches. With one-inch ore to be sampled the diameter is, of course, very much less.

This machine is one of the few that allows of a large sample being taken by making more frequent cuts. The idea of making a large sample is, of course, to insure getting a true sample from ores in which the values are irregularly distributed. If a sampler acts in such a way that it simply takes a larger slice at each cut (as in the old Brunton machine), then it is quite possible that the cut may either miss the rich spots in the stream or take an undue proportion. This is obviated in the Vezin sampler by adding extra scoops to the machine so that the same amount of ore is taken at each cut, but at very much smaller intervals. Thus, if the sampler has four scoops, each forming one-sixteenth of the circumference and makes twenty revolutions per minute, then the sample is one-fourth of the ore and a scoopful is taken at once every three-quarters of a second. If the values in the ore are more uniformly distributed two of the scoops may be taken off, and the spaces covered by blanks. Then the sample is one-eighth of the whole, and a cut is made every 1½ seconds; if only one scoop is used the sample is one-sixteenth and a cut is taken every three seconds.

Examples of Different Applications.—For chemical works crushing the ore to one twenty-fifth inch diameter, the sampler was made twenty inches high, with two scoops each occupying one-twentieth of the circumference. The weight of the sample taken was ten per cent. of the whole. With a second sampler, of identical construction, to reduce this ten per cent. the final result would be one per cent. An intermediate mixer would, of course, be required between the two machines. This size sample would work equally well on ore crushed to three-eighths inch or even one-half inch size. The samples were so arranged that a sample could be taken if required of one-quarter of one per cent. of the whole.

For coarse ore, crushed to four-inch maximum, a sampler was made to take one-fifth with one scoop. Height of machine, fifty-seven inches. Revolutions per minute, fourteen. This machine could be used for pieces six inches in diameter.

For sampling mill tailings the arrangement consists of two samplers with an intermediate mixer. Each sampler, with one scoop, takes out 1-200 of the stream, and makes if required, forty revolutions per minute, taking out a sample every 1½ seconds. The mixer is for the purpose of transforming the intermittent stream from the first sampler into a steady stream in order that the second sampler may work under proper conditions. Each machine taking 1-200 of the whole stream delivered to it, the final result is a sample one forty-thousandth of the amount of ore crushed. Therefore in a 100-ton plant (twenty-four hours) the sample made each shift of twelve hours will be 100,000÷40,000. or 2½ pounds, this pulp being contained in about five gallons of water. The height of this sampling apparatus is four feet. It could be reduced to three feet by making the mixer and the first sampler somewhat smaller, but this is advisable only when height is difficult to obtain.

Vezin Sampler.

Analysis.—The sampler is so arranged that the ore is delivered to the scoop with a velocity of about six feet per second. The designer has adopted for the speed of the scoop a velocity of about three fet per second at the center. With a machine which has a diameter of thirty inches from center to center of scoops, twenty-four revolutions per minute gives a velocity of about 3.2 feet per second, and a sample cut will be made every two and one-half seconds. The time during which the scoop is taking the sample naturally depends upon the proportion of the whole which is taken for a sample; if one-fifth, a sample will be taken during a period of 0.5 seconds. If one-tenth, during a period of 0.25 second. The shape of the cut taken is shown in the accompanying figure.

If a stream of ore amounting to twelve tons an hour be delivered to the sampler and if a regular stream be maintained, the machine will receive ore at the rate of 6.66 pounds per second. The weight of each cut taken by the sampler will then be 3.33 pounds. The weight of this cut is ample to insure that each cut will contain as nearly as possible the right proportion of coarse and fines. The machine takes 1,440 cuts an hour, and this large number of cuts insures an accurate sample even on a very spotty ore, especially as each cut has a correct shape—that of a rhombic prism. If a sample weighing one-sixteenth of the whole would contain sufficient ore to accurately represent the original parcel, the preferable arrangement would be to have two machines arranged tandem, each cutting out one-quarter.

The reasons for this arrangement are that each machine takes 1,440 cuts an hour so that a total number of cuts will be made of 2,880 in obtaining the final one-sixteenth, then. too, each cut will weigh four pounds, when taking two quarters, as against one pound when taking out one-sixteenth in one operation. When the ore is crushed coarsely, the larger weight of the cut in the tandem arrangement, aided by the increased number of cuts made, insures a more accurate sample. The consideration of the necessary weight of cut to be taken by an automatic sampler will be taken up hereafter.

Practical Results.—The machine has been used in so many places that many results have been obtained. The Metallic Extraction Company's works near Florence, erected by Philip Argall, was the first to use this machine and the results obtained at this place are interesting and valuable. The wear and tear of the machine is surpris-

ingly small. In crushing 32,000 pounds of hard Cripple Creek ore the edge of the scoops were only worn down one-half inch. The scoops were made of No. 10 B. W. G. (one-eighth inch) sheet steel. The sampling works were arranged as follows:

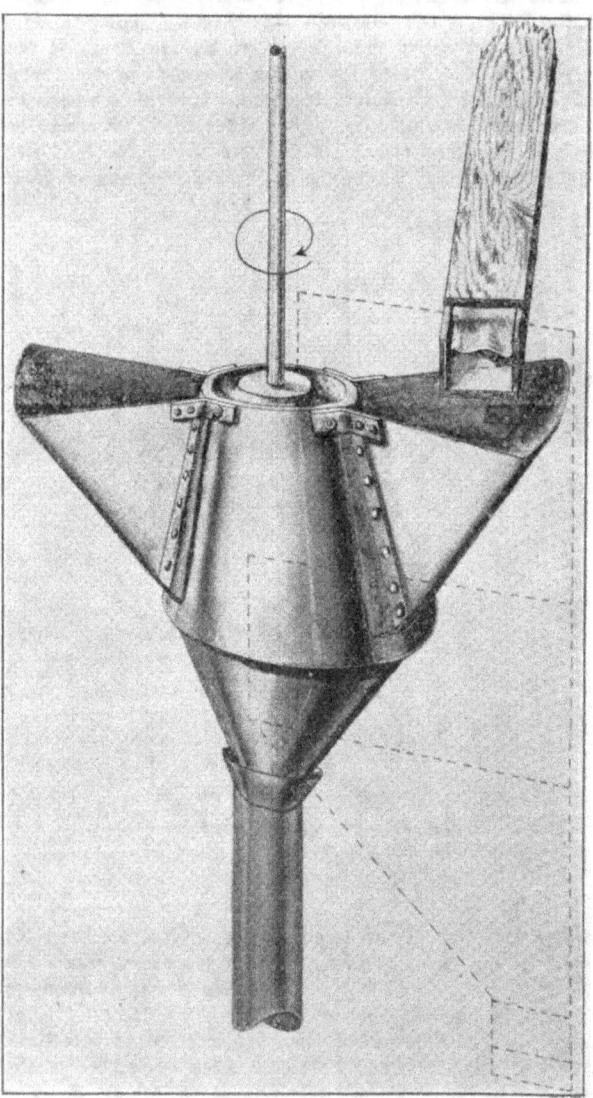

The Vezin Sampler.

The ore coming from the crushers, about one and one-half inches in diameter, was elevated and passed over the first sampler. This took out one-quarter by means of two scoops, each of which was equal to one-eighth of the periphery. The speed of the sampler was twenty revolutions per minute, so that forty cuts were taken per minute, or one every one and one-half seconds. The scoopfuls from this sampler fell into the hopper of a pair of 36x16 rolls which crushed the ore to one-quarter inch. The second elevator passed over the second sampler which took out one-tenth by means of two scoops, each of which occupied one-twentieth of the periphery. The resulting final sample was two and one-half per cent. of the original lot of ore.

The results at these works were surprisingly accurate. The ore delivered to the cyanide vats was carefully sampled and the amount of gold contained in the vat charges was compared to the amount of gold paid for. The gold purchased was 7,670.77 ounces. The gold in the vat charges was 7,661.77 ounces, a difference of only nine ounces, or $180 in a total value of gold purchased of $153,-415. On one ounce ore this would be a difference of only 3½ cents per ton. In crushing dry ore a certain amount of dust is inevitably lost, and if only two pounds of dust to the ton of ore crushed got away, this variation would be accounted for, assuming that the dust is of the same value as the ore. It is obvious, however, that the error in weighing, moisture determinations and assays would more than the ore, a loss of one pound would be sufficient.

But as the dust contains twice as much gold per ton as account for any such trivial difference.

Vezin Sampler.—The details of the construction of the machine are shown in figure 2. The claims made for the sampler for simplicity are well exemplified.

The necessary conditions that should be present for the satisfactory working of this machine are:

(1.) The stream of ore sent to the sampler should be in as solid a condition as possible and free from irregularity in its rate of flow.

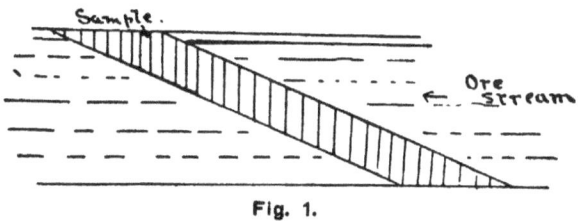

Fig. 1.

(2.) The sampler should revolve at such a rate that the velocity of the scoops at the point where they cut the center of the stream of ore should not exceed three feet per second.

(3.) If samplers are used tandem, arrangements must be made so that the ore is mixed between the samplers and sent in a steady and continuous stream to the second

sampler. This is a corrolary from condition No. 1.

If these rules be observed, and if the ore is crushed sufficiently fine, the sampler cannot fail to do good work. In order for condition No. 1 to be met, the spout leading the ore to the sampler should be at an incline of about fifty-five degrees, and should be carried to within one inch

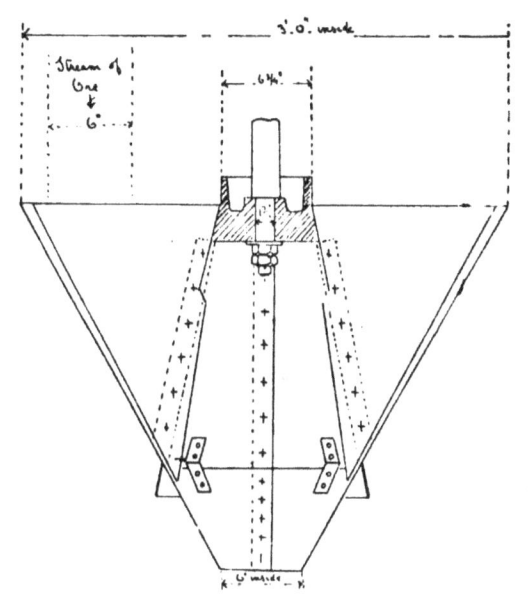

Fig 1. Section AB Fig 2

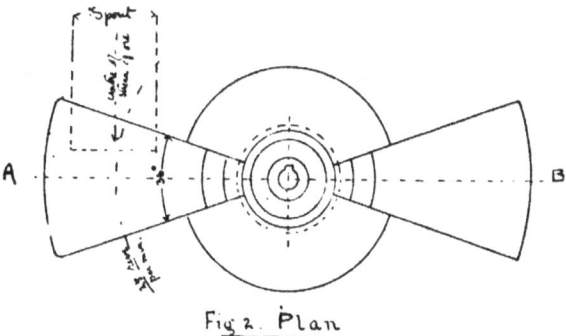

Fig 2. Plan

of the scoop. In order to prevent any pieces of ore jamming between the scoops and the spout, the cover of the latter is usually carried for the last foot of its length on hinges so that in case of jamming the spout is slightly raised and the scoop travels on without receiving any injury. The objections to a vertical spout have been considered in previous articles and it need only be recalled that such a spout delivers the ore to the sampler in a scattering stream with consequent irregularity in the work of the sampler. The speed of the machine has been limited, after trials on a working scale, to three feet per second, although the designer thinks that a speed of four feet per second would not be injurious to any marked degree.

If condition three is not observed, there is a danger of the same conditions existing as we pointed out in the Bridgeman sampler, with a consequent uncertainty in the sample cut. When very talcose or wet ores are being sampled, a dryer should be provided in order that the sampler should not get clogged.

The sampler shown in the illustration has a very large capacity, being capable of satisfactorily sampling upwards of twenty to thirty tons an hour.

It was formerly thought that the sampling of high grade ore with values irregularly distributed through the bulk of the ore, was at best merely an approximation for the correct result. This may be true for the notoriously inaccurate method of coning and quartering, but it is not true with properly designed and operated mechanical sampling works. That this is so was demonstrated at the plant of the Metallic Extraction Company near Florence. Much of the ore received at this works would assay as high as from ten to fifteen ounces ($200 to $300) per ton. Check samples corresponded within .02 ounces, or say 40 cents per ton. If, in hand sampling, on such ore, the different samples check within a dollar, remarkably good work would have been thought accomplished, and generally speaking samples on such ore would rarely agree within $2 per ton. The cost of the sampling, moreover, by machinery is but a fraction of that by hand. In order to obtain this satisfactory checking of samples by machine work, all that is needed is careful attention to such details as a common workman with ordinary intelligence can readily perform. Operating the Vezin sampler, Mr. Philip Argall,* manager of the Metallic Extraction Company's works, found that the remarkably close results already indicated could be obtained by attention to the following details.

The weight of sample required, on these high grade and London, 1902.

irregular ores, according to the size of the largest pieces of ore in the crushed material, is shown in the following table:

---

*Transactions of the Institute of Mining and Metallurgy,

Maximum Size of Cubes.	Sample, per Cent.	Pounds in 100 Tons.
1.00	20	40,000
0.25	1¼	2,500
0.625	0.0785	157
0.0171	0.005	10

In practical work, however, one would take larger quantities of the finer material, simply as a matter of extra precaution, more especially so in mills where all the ore is ultimately ground fine.

"I have used the following system quite successfully: The ore leaving the breakers had an average cube of about one inch. Twenty per cent. of this, 40,000 pounds to 100 tons, was taken as the first sample; this sample was then crushed to one-quarter-inch and ten per cent., or 4,000 pounds cut out by the second sampler and crushed to eight mesh, .0625-inch, and reduced by riffling to 250 pounds dried and crushed to about thirty mesh, .0171 and riffled down to about fifteen pounds; next passed through the sample grinder, reducing it to, say, 90-100 mesh and riffling this down to one pound, which was ground on the buckboard to pass 120 mesh sieve, .004-inch, and put up in four envelopes, one for the seller, one for the purchaser, one sealed by both parties for umpire, and the fourth simply held as a reserve. If this work is well done a one-half assay ton from any of the 120-mesh pulps would check within .02 ounces or, say, 40 cents per ton.

"At first I only crushed the samples to 100 mesh, and often found difficulties in getting duplicates to check. Since using the 120 mesh, however, and taking care to have a thorough mixing of the sample at the various stages of the process, there has seldom, if ever, been any material difference in the original and duplicates between the four samples cut out of the final pulp."

Cost of Sampling.—According to Mr. Argall, the cost of sampling at the Metallic Extraction Company's works was as follows:

Labor account	$0.04325
Operating expenses	0.00950
Maintenance	0.05420
Power	0.01457
Total cost per ton	$0.12152

The power required to operate the entire sampling works at the same place was 70 H. P.

CHAPTER XIV.

The accuracy of any sample taken by a machine depends, very largely, upon the sample containing the same proportion of fines and coarse as the original lot of ore. For public sampling works, where all kinds of ore are being received, the sampling devices used must give such a sample. Let us take a case. We have seen sampling machines "tested" by running crushed copper matte through them. The difference, in value, between the coarse and fines is

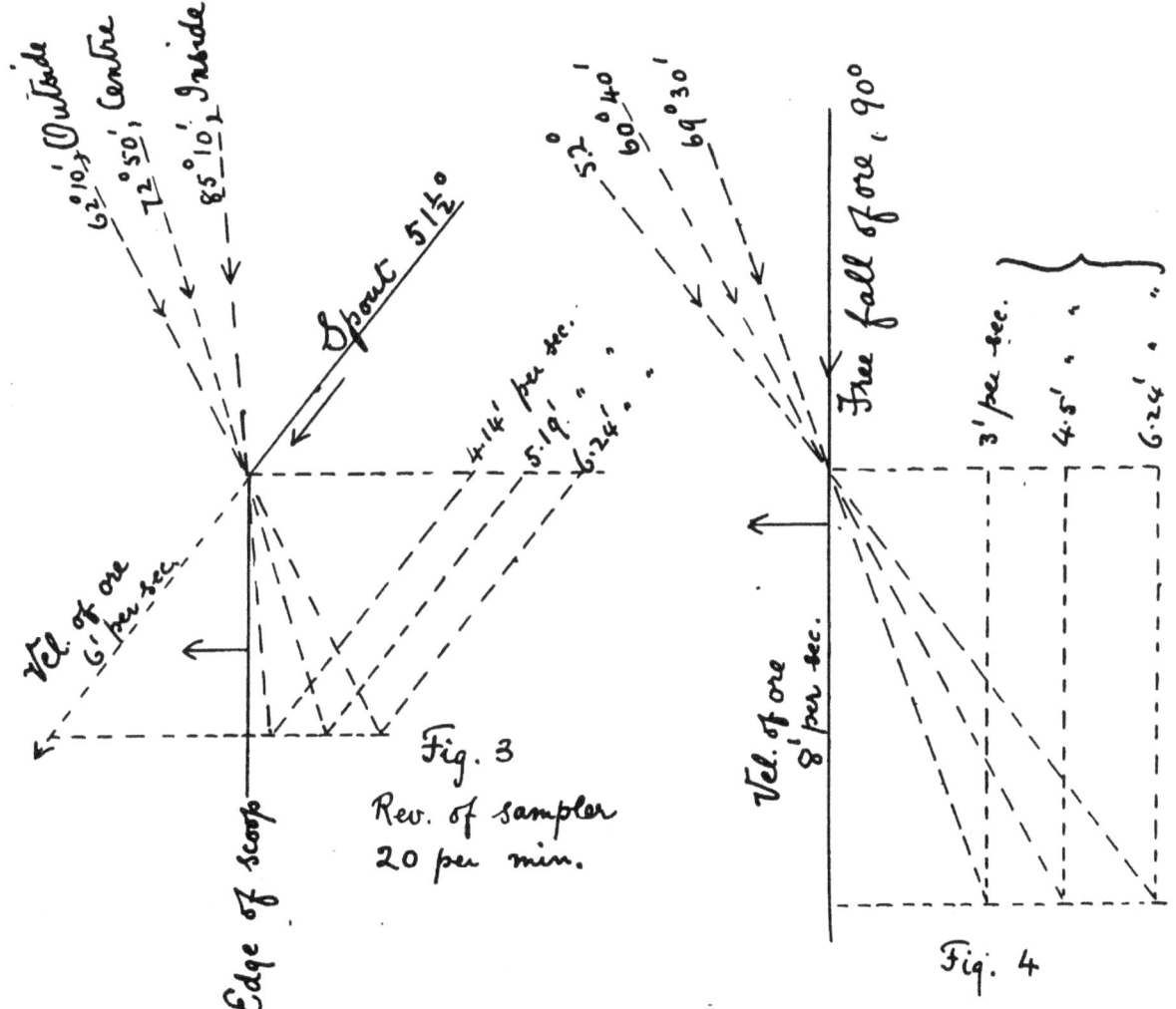

Fig. 3
Rev. of sampler
20 per min.

Fig. 4

little or nothing in each case. If a cut is made at frequent enough intervals almost any machine would give correct samples on this material. But if such ores as those of Cripple Creek district, or of Clear Creek county, Colorado, were being put through, it is essential that the sample should have the same proportion of coarse and fines as the original ore. If we have a Cripple Creek ore, crushed to a two-inch ring, going through the machine and a screen analysis gave the following results:

Ore on 1 inch.............. 25%	Assay	$ 22.00
Ore on ½ inch............. 40%	Assay	90.00
Ore on ⅛ inch............. 22%	Assay	115.00
Ore through ⅛ inch....... 13%	Assay	200.00
100%		$ 92.80

Then should the machine give a sample containing only 20% of material remaining on a 1-inch screen the sample

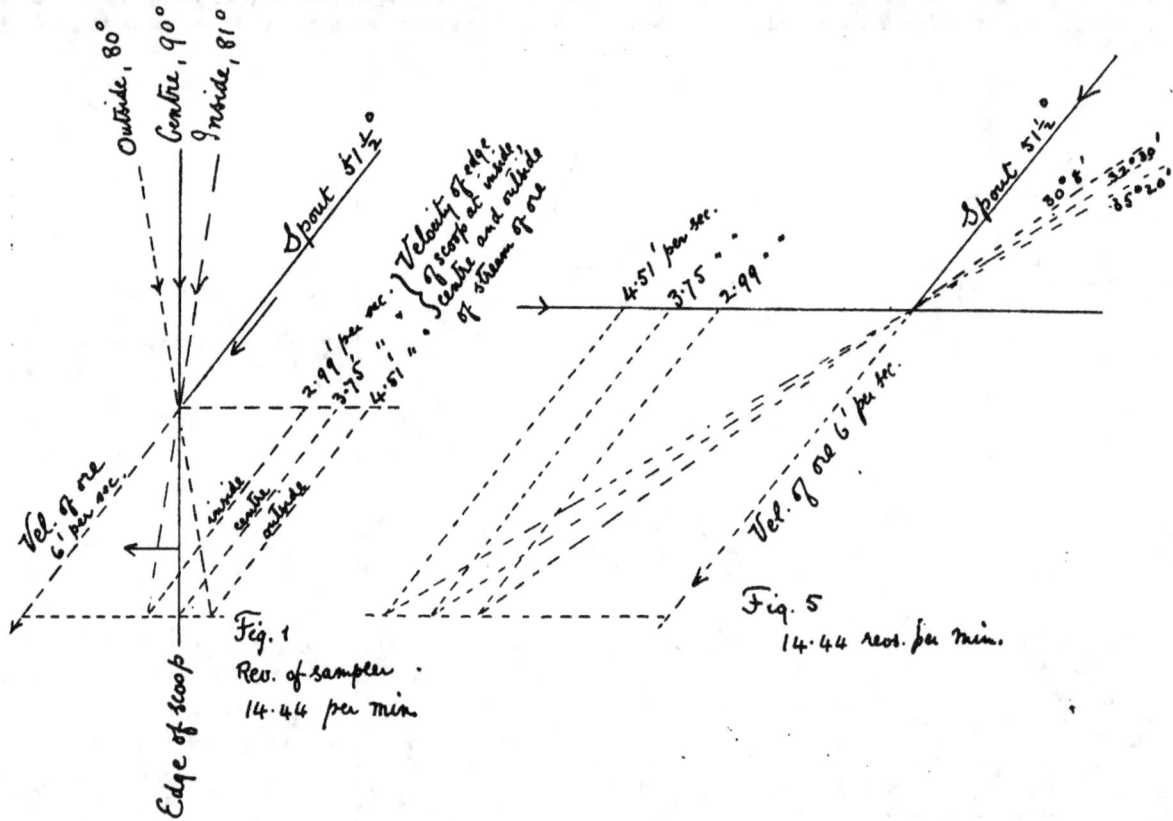

Fig. 1
Rev. of sampler.
14.44 per min.

Fig. 5
14.44 revs. per min.

would assay about $97.50, or $4.70 per ton too high—an error far in excess of what would be allowed in good modern practice.

We can deduce from these facts (1) That a sampling machine, in order to do good work, should fulfil the conditions laid down above, and (2) that, in order for a test of a sampling machine to be complete, it should include a set of screen analyses besides assays. It goes without saying, to those who have followed the foregoing articles, that the sample of the rejected ore should be made by the method of taking alternate shovelfuls. We lay great stress

----

*According to Mr. Argall, manager of the Metallic Extraction Company's mill at Cyanide, Colorado, it makes no difference in what direction the ore falls on the edge of the scoop. Mr. Argall had two samplers, so arranged as to take duplicate samples, but cutting the stream in opposite directions. The sellers of the ore had the choice of the two samples thus made, but it was found that the difference, in value, between the samples was nil. The duplicate samples were only made during a period of a few months, or until it was clearly demonstrated that there was no difference in value. Thereafter duplicate samples were only made when called for by the ore seller.

upon this method of testing since it demonstrates, in conjunction with the assays, not only whether a machine is capable of taking a good sample, but also another very important point, viz, whether the cuts are being taken often enough. For example, if we find that the screen analyses of the sample and of the rejected ore are the same, or practically so, but that the assays do not agree as closely as they should, then we see no other possible explanation than that the cut has not been taken frequently enough and hence a long portion of rich or poor ore, as the case might be, has got through the machine without being sampled. A machine that is perfectly good in principle might be condemned if the method of checking by assays only is adopted. The fault might be that the cuts are taken too far apart.

Some very ingenious hypotheses have been advanced in connection with the point as to whether a machine takes enough fines or coarse. A favorite idea appears to be that a coarse piece of ore falling at an angle on the edge of the scoop might bounce off so far as to escape being caught. Some hold that the ore should fall vertically, others that if the ore fell at an angle in a direction opposite to the line of travel of the scoop a sample too high in value would be obtained owing to the machine not taking enough of the coarse ore. On the other hand if the ore falls at an angle in the same direction as the line of travel of the scoop then the sample is too low, owing to the machine taking too much of the coarse ore. One genius laid down a law to the effect that a vertical fall, if of sufficient height, gave a sample which was too low, and that an inclined spout gave a sample which was too high, and hence if a knee-shaped spout were constructed with the vertical part just one foot in length the errors would balance, and a correct sample would be the result! We have made many inquiries as to the effect of the direction of fall upon the value of the sample obtained, and, so far as we can learn, it appears to have no effect on the reliability of the sample.* In designing a sampling plant, however, it is as well to so arrange the spouts and the velocity of the scoops that the ore may fall vertically on the edge of the scoop.

The diagrams shown herewith will make the matter plain: Fig. 1 shows the angle at which the falling stream of ore strikes the edge of the scoop. It will be seen that when the ore falls at an angle of 51½° the result is that the ore falls on the moving scoop at an angle of from 80° (+) to 80° (—) with an average of about 90°. When a piece of ore falls vertically on a sheet metal edge it will simply fall on one side or the other according as the center of gravity of the piece falls; there is no question of bouncing off according to an angle of reflection. If there is any bounce it will be more or less vertical in direction and there can be no question as to large pieces of ore jumping off in a horizontal direction and hence escaping being caught by the scoop. In the Vezin sampler, sixty inches diameter, with 14.44 revolutions per minute, spout inclined 51½°, the ore particles having a velocity of six feet per second, the ore will fall vertically on the edge of the scoop (see fig. 1). If the sampler revolves at twenty revolutions per minute the ore will strike at a flatter angle (see fig. 3). If the ore falls vertically then with the scoop making 14.44 revolutions per minute the ore will strike the edge of the scoop at quite a flat angle (see fig. 4). With the ore falling in an opposite direction to the travel of the scoop then the ore will strike at a very flat angle indeed (see fig. 5). Paradoxical as it may appear to some, in order to obtain a vertical fall of ore upon a moving edge an inclined spout is required and not a vertical one.

CHAPTER XV.

In the foregoing chapters the main consideration has been to deal with principles which underlie the satisfactory sampling of ore. In the remaining notes we will give some examples of the manner in which the whole process is carried on at various places.

The mill to be described in this note is that of the Park City, Utah, sampling works, commonly known as "The Mackintosh Mill." This plant is a very recent construction and is designed to replace the old works which have been sampling from 10,000 to 11,000 tons of ore a month for many years. The sampler handles all the shipping ores of the Park City district with the exception of the ores produced by the Silver King Mining Company, which has its own sampling works.

The Mackintosh sampler, being a custom works, with most of its output handled by smelters, the ore has to be kept as coarse as possible. Owing to the low treatment charge it is necessary that the plant be as automatic as possible.

In carrying out the operations the ore brought to the works in railroad cars is taken by a twenty-four-inch belt conveyor to a No. 5 Gates crusher. From this crusher the ore is elevated and delivered to a Vezin sampler, which cuts out one-fifth, which is retained in the sampling system, and four-fifths are rejected and go by way of an elevator to the shipping bin. The sample cuts are sent to a set of 36x14-inch rolls, which reduce the ore to about one inch. The ore is elevated to a second Vezin sampler, which again cuts out one-fifth, and rejects four-fifths, which go as before to the shipping bin. The sample cuts which are now 1-25th of the whole, go to a second set of rolls, 30x10 inches. The ore gets a second nip, and the reduced product goes to a third sampler, which cuts out one-fifth, as before, making the sample at this point 1-125th of the whole. This sample is then sent to a set of 24x8-inch rolls and again crushed. The crushed material is sent to a fourth sampler, one-fifth again cut out, yielding a sample of 1-625th of the whole, the rejected ore, as before, going to the shipping bin. The sample cuts are then sent to a set of 9x4-inch rolls and are crushed again, making about 1-10th inch product, which goes to the fifth, or last, sampler, which cuts out one-fifth, which now represents 1-3125th of the whole amount of ore sampled. The rejected ore goes to the shipping bin, and the sample is then taken to a fine sample grinder, which reduces the ore to a

50-mesh screen, and is cut down by the ordinary methods for the assayer.

The power plant consists of two Scotch marine boilers of 145-H.P., with automatic stokers; the engine is a 16x36 Allis-Corliss. The sampler is designed to have a capacity of 500 tons daily. The building is lit by electric light and steam heated. The plant is also provided with self-registering car scales.

These works seem admirably designed to furnish reliable samples by an entirely automatic system. We would question, however, whether it would be necessary to have quite as many Vezin samplers, and we think that the last sampler might be advantageously replaced by a Jones riffle sampler. This, however, is a matter of individual judgment. The plant seems to be in advance on anything at present erected in the West, and reflects much credit upon the management of the Mackintosh estate.

In considering this plant we may offer a few words of friendly criticism. It is probable that both the ore buyers and sellers would insist on the elevators being cleaned out before sampling a new lot of ore. Even with drop bottoms in the elevator boots, this involves annoying delays. The objection might also be urged that the fourth and fifth samplers have very little to do, i. e., that their capacity is very much greater than required. Thus, in a 100-ton lot of ore the amount that passes over No. 4 sampler is 1,600 pounds, which it is assumed has been crushed to one-quarter-inch in diameter. If the stream were but ½x2 inches, flowing at six feet per second, the amount would be 1-24th cubic foot per second, and, assuming the crushed ore weighs 100 pounds per cubic foot, the 1,600 pounds would pass in sixty-four seconds. The capacity of the sampler with this small stream of one square inch cross-section would be 15,000 pounds, or 7½ tons per hour. A sampler, seven inches in diameter at the center of the stream of ore, with two scoops, each 2.2 inches wide, making ninety revolutions per minute, would be sufficiently large. If preferred it could be made fourteen inches in diameter with two scoops 4.4 inches wide, or with four scoops 2.2 inches wide. It should make forty-five revolutions per minute.

It might be suggested that instead of samplers 4 and 5, with their elevators, a platform hoist and a Jones riffle sampler might be substituted. There would be no cleaning out of elevators and the arrangement would be rather

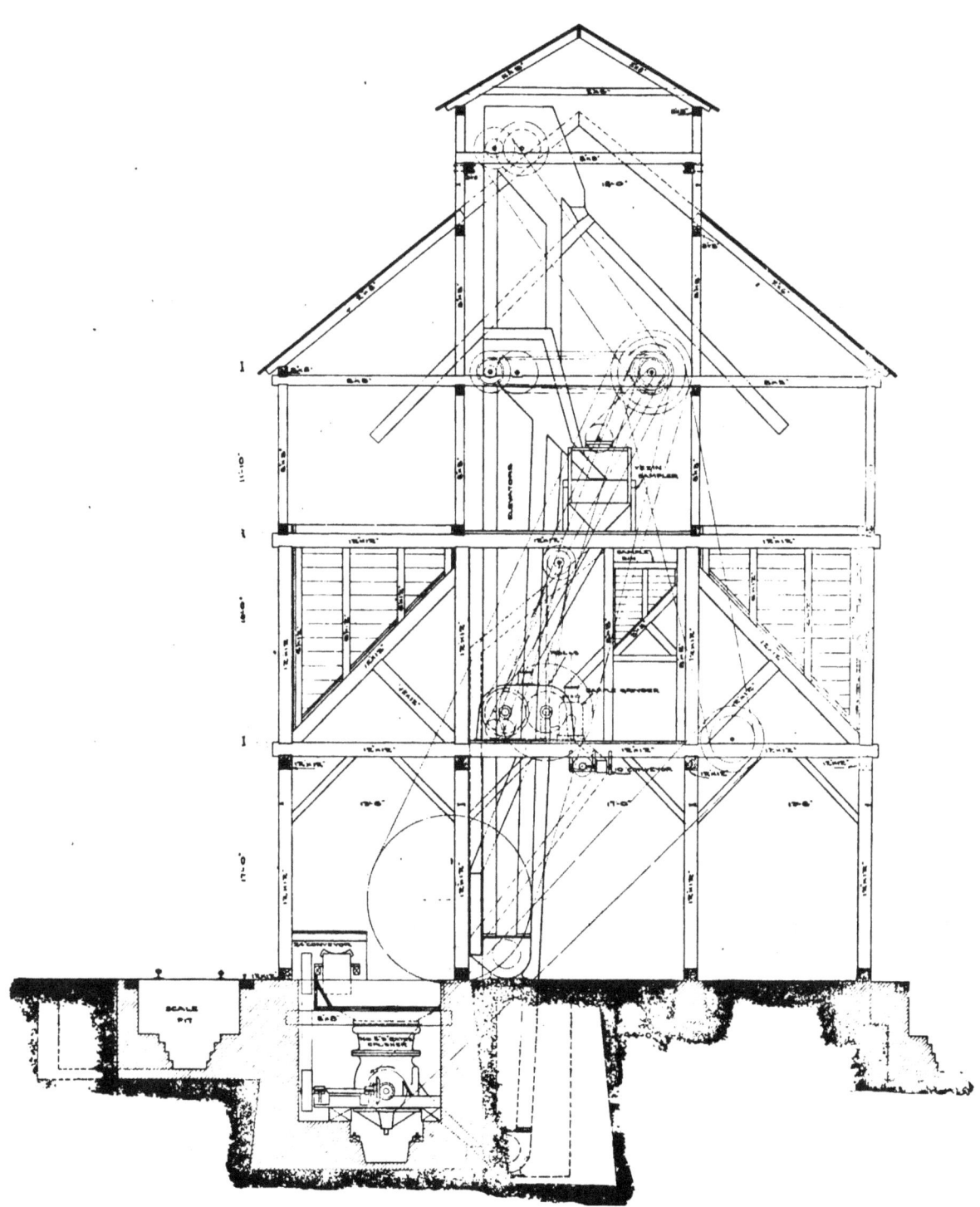

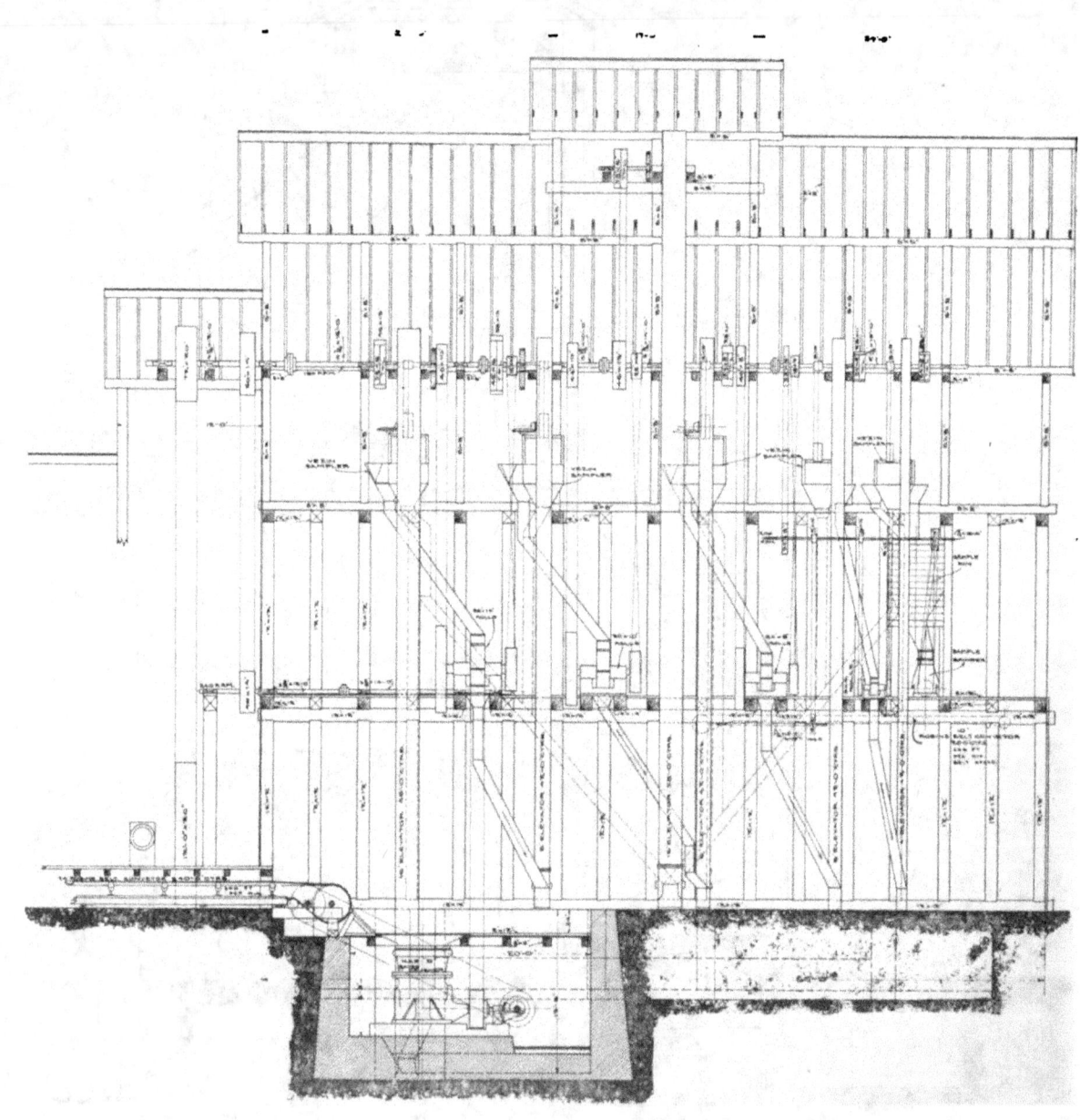

simpler and there would be two less elevators to clean. The two-wheel scoop barrow could be used, made to hold 800 or 1,600 pounds, as preferred. The Jones riffle sampler has been arranged in a series of three so that the resulting sample is one-eighth. If a single one is used the subdivision would take longer, but has one advantage that inspires confidence, as there is a complete mixing between divisions. The ordinary scoop used with an ordinary riffle sampler holds about sixteen pounds of ore, so that 1,600 pounds represent 100 scoopfuls. If there are twelve riffles in the single Jones sampler pouring 1,600 pounds once may be said to represent 1,200 cuts.

The platform hoist, of course, serves to raise the material to the feeding hopper of the smaller rolls.

The arrangement of this sampling works is so very good that it seems almost like hypercriticism to suggest that anything else might be preferred.

The use of the riffle sampler offers a very good opportunity for saving a duplicate to be held in reserve to guard against an accident to the original.

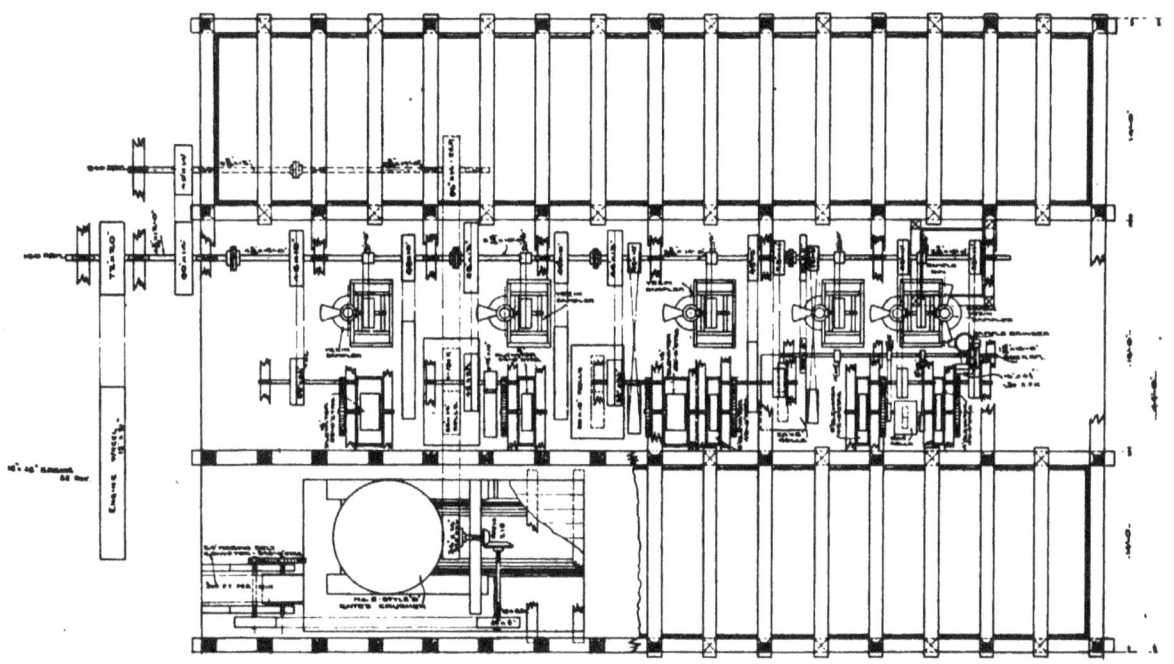

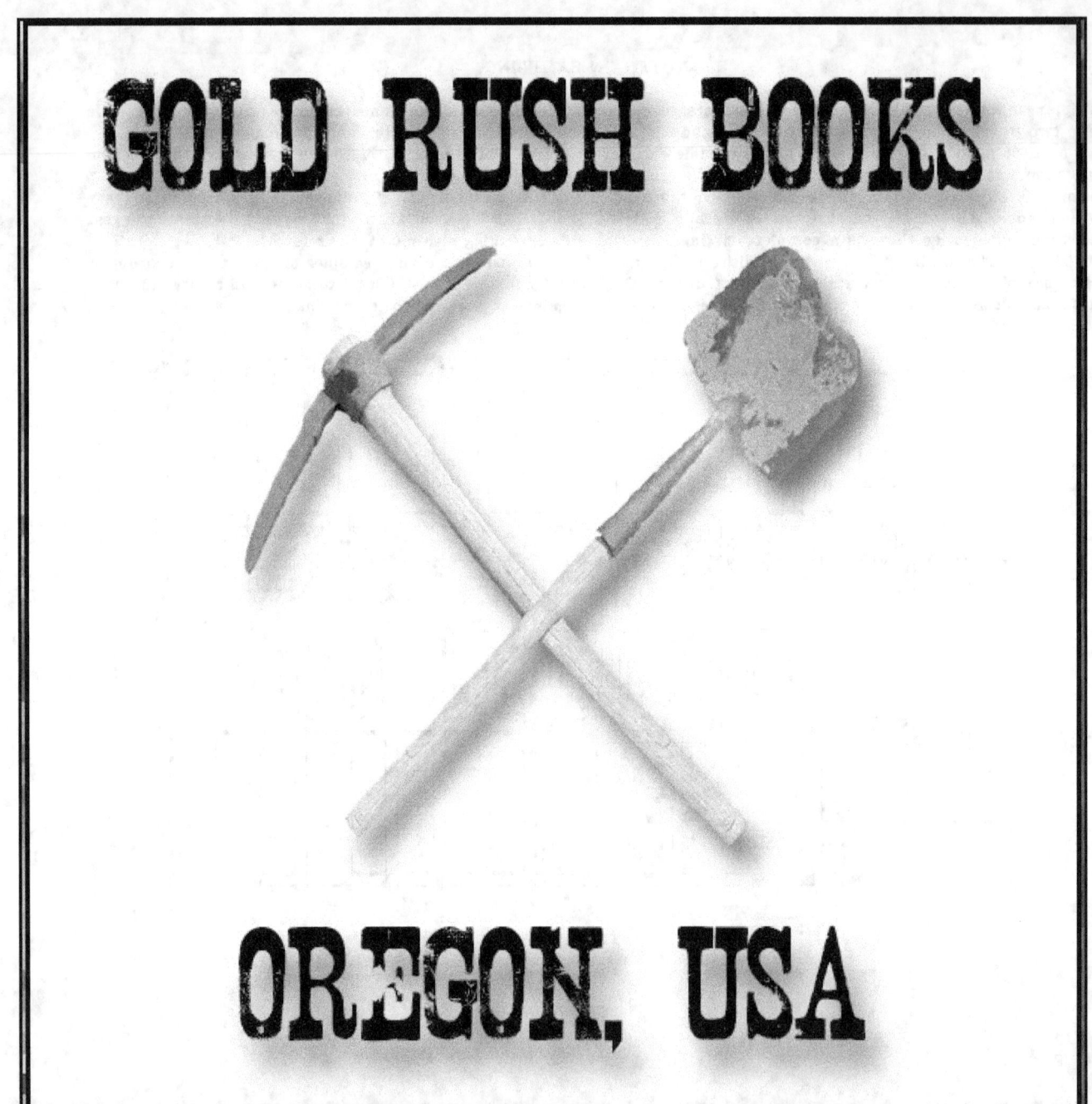

GOLD RUSH BOOKS

OREGON, USA

www.GoldMiningBooks.com

Books On Mining

Visit: www.goldminingbooks.com to order your copies or ask your favorite book seller to offer them.

Mining Books by Kerby Jackson

Gold Dust: Stories From Oregon's Mining Years - Oregon mining historian and prospector, Kerby Jackson, brings you a treasure trove of seventeen stories on Southern Oregon's rich history of gold prospecting, the prospectors and their discoveries, and the breathtaking areas they settled in and made homes. 5" X 8", 98 ppgs. Retail Price: $11.99

The Golden Trail: More Stories From Oregon's Mining Years - In his follow-up to "Gold Dust: Stories of Oregon's Mining Years", this time around, Jackson brings us twelve tales from Oregon's Gold Rush, including the story about the first gold strike on Canyon Creek in Grant County, about the old timers who found gold by the pail full at the Victor Mine near Galice, how Iradel Bray discovered a rich ledge of gold on the Coquille River during the height of the Rogue River War, a tale of two elderly miners on the hunt for a lost mine in the Cascade Mountains, details about the discovery of the famous Armstrong Nugget and others. 5" X 8", 70 ppgs. Retail Price: $10.99

Oregon Mining Books

Geology and Mineral Resources of Josephine County, Oregon - Unavailable since the 1970's, this important publication was originally compiled by the Oregon Department of Geology and Mineral Industries and includes important details on the economic geology and mineral resources of this important mining area in South Western Oregon. Included are notes on the history, geology and development of important mines, as well as insights into the mining of gold, copper, nickel, limestone, chromium and other minerals found in large quantities in Josephine County, Oregon. 8.5" X 11", 54 ppgs. Retail Price: $9.99

Mines and Prospects of the Mount Reuben Mining District - Unavailable since 1947, this important publication was originally compiled by geologist Elton Youngberg of the Oregon Department of Geology and Mineral Industries and includes detailed descriptions, histories and the geology of the Mount Reuben Mining District in Josephine County, Oregon. Included are notes on the history, geology, development and assay statistics, as well as underground maps of all the major mines and prospects in the vicinity of this much neglected mining district. 8.5" X 11", 48 ppgs. Retail Price: $9.99

The Granite Mining District - Notes on the history, geology and development of important mines in the well known Granite Mining District which is located in Grant County, Oregon. Some of the mines discussed include the Ajax, Blue Ribbon, Buffalo, Continental, Cougar-Independence, Magnolia, New York, Standard and the Tillicum. Also included are many rare maps pertaining to the mines in the area. 8.5" X 11", 48 ppgs. Retail Price: $9.99

Ore Deposits of the Takilma and Waldo Mining Districts of Josephine County, Oregon - The Waldo and Takilma mining districts are most notable for the fact that the earliest large scale mining of placer gold and copper in Oregon took place in these two areas. Included are details about some of the earliest large gold mines in the state such as the Llano de Oro, High Gravel, Cameron, Platerica, Deep Gravel and others, as well as copper mines such as the famous Queen of Bronze mine, the Waldo, Lily and Cowboy mines. This volume also includes six maps and 20 original illustrations. 8.5" X 11", 74 ppgs. Retail Price: $9.99

Metal Mines of Douglas, Coos and Curry Counties, Oregon - Oregon mining historian Kerby Jackson introduces us to a classic work on Oregon's mining history in this important re-issue of Bulletin 14C Volume 1, otherwise known as the Douglas, Coos & Curry Counties, Oregon Metal Mines Handbook. Unavailable since 1940, this important publication was originally compiled by the Oregon Department of Geology and Mineral Industries includes detailed descriptions, histories and the geology of over 250 metallic mineral mines and prospects in this rugged area of South West Oregon. 8.5" X 11", 158 ppgs. Retail Price: $19.99

Metal Mines of Jackson County, Oregon - Unavailable since 1943, this important publication was originally compiled by the Oregon Department of Geology and Mineral Industries includes detailed descriptions, histories and the geology of over 450 metallic mineral mines and prospects in Jackson County, Oregon. Included are such famous gold mining areas as Gold Hill, Jacksonville, Sterling and the Upper Applegate. **8.5" X 11", 220 ppgs. Retail Price: $24.99**

Metal Mines of Josephine County, Oregon - Oregon mining historian Kerby Jackson introduces us to a classic work on Oregon's mining history in this important re-issue of Bulletin 14C, otherwise known as the Josephine County, Oregon Metal Mines Handbook. Unavailable since 1952, this important publication was originally compiled by the Oregon Department of Geology and Mineral Industries includes detailed descriptions, histories and the geology of over 500 metallic mineral mines and prospects in Josephine County, Oregon. **8.5" X 11", 250 ppgs. Retail Price: $24.99**

Metal Mines of North East Oregon - Oregon mining historian Kerby Jackson introduces us to a classic work on Oregon's mining history in this important re-issue of Bulletin 14A and 14B, otherwise known as the North East Oregon Metal Mines Handbook. Unavailable since 1941, this important publication was originally compiled by the Oregon Department of Geology and Mineral Industries and includes detailed descriptions, histories and the geology of over 750 metallic mineral mines and prospects in North Eastern Oregon. **8.5" X 11", 310 ppgs. Retail Price: $29.99**

Metal Mines of North West Oregon - Oregon mining historian Kerby Jackson introduces us to a classic work on Oregon's mining history in this important re-issue of Bulletin 14D, otherwise known as the North West Oregon Metal Mines Handbook. Unavailable since 1951, this important publication was originally compiled by the Oregon Department of Geology and Mineral Industries and includes detailed descriptions, histories and the geology of over 250 metallic mineral mines and prospects in North Western Oregon. **8.5" X 11", 182 ppgs. Retail Price: $19.99**

Mines and Prospects of Oregon - Mining historian Kerby Jackson introduces us to a classic mining work by the Oregon Bureau of Mines in this important re-issue of The Handbook of Mines and Prospects of Oregon. Unavailable since 1916, this publication includes important insights into hundreds of gold, silver, copper, coal, limestone and other mines that operated in the State of Oregon around the turn of the 19th Century. Included are not only geological details on early mines throughout Oregon, but also insights into their history, production, locations and in some cases, also included are rare maps of their underground workings. **8.5" X 11", 314 ppgs. Retail Price: $24.99**

Lode Gold of the Klamath Mountains of Northern California and South West Oregon
(See California Mining Books)

Mineral Resources of South West Oregon - Unavailable since 1914, this publication includes important insights into dozens of mines that once operated in South West Oregon, including the famous gold fields of Josephine and Jackson Counties, as well as the Coal Mines of Coos County. Included are not only geological details on early mines throughout South West Oregon, but also insights into their history, production and locations. **8.5" X 11", 154 ppgs. Retail Price: $11.99**

Chromite Mining in The Klamath Mountains of California and Oregon
(See California Mining Books)

Southern Oregon Mineral Wealth - Unavailable since 1904, this rare publication provides a unique snapshot into the mines that were operating in the area at the time. Included are not only geological details on early mines throughout South West Oregon, but also insights into their history, production and locations. Some of the mining areas include Grave Creek, Greenback, Wolf Creek, Jump Off Joe Creek, Granite Hill, Galice, Mount Reuben, Gold Hill, Galls Creek, Kane Creek, Sardine Creek, Birdseye Creek, Evans Creek, Foots Creek, Jacksonville, Ashland, the Applegate River, Waldo, Kerby and the Illinois River, Althouse and Sucker Creek, as well as insights into local copper mining and other topics. **8.5" X 11", 64 ppgs. Retail Price: $8.99**

Geology and Ore Deposits of the Takilma and Waldo Mining Districts - Unavailable since the 1933, this publication was originally compiled by the United States Geological Survey and includes details on gold and copper mining in the Takilma and Waldo Districts of Josephine County, Oregon. The Waldo and Takilma mining districts are most notable for the fact that the earliest large scale mining of placer gold and copper in Oregon took place in these two areas. Included in this report are details about some of the earliest large gold mines in the state such as the Llano de Oro, High Gravel, Cameron, Platerica, Deep Gravel and others, as well as copper mines such as the famous Queen of Bronze mine, the Waldo, Lily and Cowboy mines. In addition to geological examinations, insights are also provided into the production, day to day operations and early histories of these mines, as well as calculations of known mineral reserves in the area. This volume also includes six maps and 20 original illustrations. **8.5" X 11", 74 ppgs. Retail Price: $9.99**

Gold Mines of Oregon - Oregon mining historian Kerby Jackson introduces us to a classic work on Oregon's mining history in this important re-issue of Bulletin 61, otherwise known as "Gold and Silver In Oregon". Unavailable since 1968, this important publication was originally compiled by geologists Howard C. Brooks and Len Ramp of the Oregon Department of Geology and Mineral Industries and includes detailed descriptions, histories and the geology of over 450 gold mines Oregon. Included are notes on the history, geology and gold production statistics of all the major mining areas in Oregon including the Klamath Mountains, the Blue Mountains and the North Cascades. While gold is where you find it, as every miner knows, the path to success is to prospect for gold where it was previously found. 8.5" X 11", 344 ppgs. **Retail Price: $24.99**

Mines and Mineral Resources of Curry County Oregon - Originally published in 1916, this important publication on Oregon Mining has not been available for nearly a century. Included are rare insights into the history, production and locations of dozens of gold mines in Curry County, Oregon, as well as detailed information on important Oregon mining districts in that area such as those at Agness, Bald Face Creek, Mule Creek, Boulder Creek, China Diggings, Collier Creek, Elk River, Gold Beach, Rock Creek, Sixes River and elsewhere. Particular attention is especially paid to the famous beach gold deposits of this portion of the Oregon Coast. 8.5" X 11", 140 ppgs. **Retail Price: $11.99**

Chromite Mining in South West Oregon - Originally published in 1961, this important publication on Oregon Mining has not been available for nearly a century. Included are rare insights into the history, production and locations of nearly 300 chromite mines in South Western Oregon. 8.5" X 11", 184 ppgs. **Retail Price: $14.99**

Mineral Resources of Douglas County Oregon - Originally published in 1972, this important publication on Oregon Mining has not been available for nearly forty years. Included are rare insights into the geology, history, production and locations of numerous gold mines and other mining properties in Douglas County, Oregon. 8.5" X 11", 124 ppgs. **Retail Price: $11.99**

Mineral Resources of Coos County Oregon - Originally published in 1972, this important publication on Oregon Mining has not been available for nearly forty years. Included are rare insights into the geology, history, production and locations of numerous gold mines and other mining properties in Coos County, Oregon. 8.5" X 11", 100 ppgs. **Retail Price: $11.99**

Mineral Resources of Lane County Oregon - Originally published in 1938, this important publication on Oregon Mining has not been available for nearly seventy five years. Included are extremely rare insights into the geology and mines of Lane County, Oregon, in particular in the Bohemia, Blue River, Oakridge, Black Butte and Winberry Mining Districts. 8.5" X 11", 82 ppgs. **Retail Price: $9.99**

Mineral Resources of the Upper Chetco River of Oregon: Including the Kalmiopsis Wilderness - Originally published in 1975, this important publication on Oregon Mining has not been available for nearly forty years. Withdrawn under the 1872 Mining Act since 1984, real insight into the minerals resources and mines of the Upper Chetco River has long been unavailable due to the remoteness of the area. Despite this, the decades of battle between property owners and environmental extremists over the last private mining inholding in the area has continued to pique the interest of those interested in mining and other forms of natural resource use. Gold mining began in the area in the 1850's and has a rich history in this geographic area, even if the facts surrounding it are little known. Included are twenty two rare photographs, as well as insights into the Becca and Morning Mine, the Emmly Mine (also known as Emily Camp), the Frazier Mine, the Golden Dream or Higgins Mine, Hustis Mine, Peck Mine and others. 8.5" X 11", 64 ppgs. **Retail Price: $8.99**

Gold Dredging in Oregon - Originally published in 1939, this important publication on Oregon Mining has not been available for nearly seventy five years. Included are extremely rare insights into the history and day to day operations of the dragline and bucketline gold dredges that once worked the placer gold fields of South West and North East Oregon in decades gone by. Also included are details into the areas that were worked by gold dredges in Josephine, Jackson, Baker and Grant counties, as well as the economic factors that impacted this mining method. This volume also offers a unique look into the values of river bottom land in relation to both farming and mining, in how farm lands were mined, re-soiled and reclamated after the dredges worked them. Featured are hard to find maps of the gold dredge fields, as well as rare photographs from a bygone era. 8.5" X 11", 86 ppgs. **Retail Price: $8.99**

Quick Silver Mining in Oregon - Originally published in 1963, this important publication on Oregon Mining has not been available for over fifty years. This publication includes details into the history and production of Elemental Mercury or Quicksilver in the State of Oregon. 8.5" X 11", 238 ppgs. **Retail Price: $15.99**

Mines of the Greenhorn Mining District of Grant County Oregon - Originally published in 1948, this important publication on Oregon Mining has not been available for over sixty five years. In this publication are rare insights into the mines of the famous Greenhorn Mining District of Grant County, Oregon, especially the famous Morning Mine. Also included are details on the Tempest, Tiger, Bi-Metallic, Windsor, Psyche, Big Johnny, Snow Creek, Banzette and Paramount Mines, as well as prospects in the vicinities in the famous mining areas of Mormon Basin, Vinegar Basin and Desolation Creek. Included are hard to find mine maps and dozens of rare photographs from the bygone era of Grant County's rich mining history. 8.5" X 11", 72 ppgs. **Retail Price: $9.99**

Geology of the Wallowa Mountains of Oregon: Part I (Volume 1) - Originally published in 1938, this important publication on Oregon Mining has not been available for nearly seventy five years. Included are details on the geology of this unique portion of North Eastern Oregon. This is the first part of a two book series on the area. Accompanying the text are rare photographs and historic maps.8.5" X 11", **92 ppgs. Retail Price: $9.99**

Geology of the Wallowa Mountains of Oregon: Part II (Volume 2) - Originally published in 1938, this important publication on Oregon Mining has not been available for nearly seventy five years. Included are details on the geology of this unique portion of North Eastern Oregon. This is the first part of a two book series on the area. Accompanying the text are rare photographs and historic maps.8.5" X 11", **94 ppgs. Retail Price: $9.99**

Field Identification of Minerals For Oregon Prospectors - Originally published in 1940, this important publication on Oregon Mining has not been available for nearly seventy five years. Included in this volume is an easy system for testing and identifying a wide range of minerals that might be found by prospectors, geologists and rockhounds in the State of Oregon, as well as in other locales. Topics include how to put together your own field testing kit and how to conduct rudimentary tests in the field. This volume is written in a clear and concise way to make it useful even for beginners. **8.5" X 11", 158 ppgs. Retail Price: $14.99**

The Bohemia Mining District of Oregon - Originally published in 1900, this important publication on Oregon Mining has not been available for over a century. Included in this volume are important insights into the famous Bohemia Mining District of Oregon, including the histories and locations of important gold mines in the area such as the Ophir Mine, Clarence, Acturas, Peek-a-boo, White Swan, Combination Mine, the Musick Mine, The California, White Ghost, The Mystery, Wall Street, Vesuvius, Story, Lizzie Bullock, Delta, Elsie Dora, Golden Slipper, Broadway, Champion Mine, Knott, Noonday, Helena, White Wings, Riverside and others. Also included are notes on the nearby Blue River Mining District. **8.5" X 11", 58 ppgs. Retail Price: $9.99**

The Gold Fields of Eastern Oregon - Unavailable since 1900, this publication was originally compiled by the Baker City Chamber of Commerce Offering important insights into the gold mining history of Eastern Oregon, "The Gold Fields of Eastern Oregon" sheds a rare light on many of the gold mines that were operating at the turn of the 19th Century in Baker County and Grant County in North Eastern Oregon. Some of the areas featured include the Cable Cove District, Baisely-Elhorn, Granite, Red Boy, Bonanza, Susanville, Sparta, Virtue, Vaughn, Sumpter, Burnt River, Rye Valley and other mining districts. Included is basic information on not only many gold mines that are well known to those interested in Eastern Oregon mining history, but also many mines and prospects which have been mostly lost to the passage of time. Accompanying are numerous rare photos **8.5" X 11", 78 ppgs. Retail Price: $10.99**

Gold Mining in Eastern Oregon - Originally published in 1938, this important publication on Oregon Mining has not been available for over a century. Included in this volume are important insights into the famous mining districts of Eastern Oregon during the late 1930's. Particular attention is given to those gold mines with milling and concentrating facilities in the Greenhorn, Red Boy, Alamo, Bonanza, Granite, Cable Cove, Cracker Creek, Virtue, Keating, Medical Springs, Sanger, Sparta, Chicken Creek, Mormon Basin, Connor Creek, Cornucopia and the Bull Run Mining Districts. Some of the mines featured include the Ben Harrison, North Pole-Columbia, Highland Maxwell, Baisley-Elkhorn, White Swan, Balm Creek, Twin Baby, Gem of Sparta, New Deal, Gleason, Gifford-Johnson, Cornucopia, Record, Bull Run, Orion and others. Of particular interest are the mill flow sheets and descriptions of milling operations of these mines. **8.5" X 11", 68 ppgs. Retail Price: $8.99**

The Gold Belt of the Blue Mountains of Oregon - Originally published in 1901, this important publication on Oregon Mining has not been available for over a century. Included in this volume are rare insights into the gold deposits of the Blue Mountains of North East Oregon, including the history of their early discovery and early production. Extensive details are offered on this important mining area's mineralogy and economic geology, as well as insights into nearby gold placers, silver deposits and copper deposits. Featured are the Elkhorn and Rock Creek mining districts, the Pocahontas district, Auburn and Minersville districts, Sumpter and Cracker Creek, Cable Cove, the Camp Carson district, Granite, Alamo, Greenhorn, Robinsonville, the Upper Burnt River Valley and Bonanza districts, Susanville, Quartzburg, Canyon Creek, Virtue, the Copper Butte district, the North Powder River, Sparta, Eagle Creek, Cornucopia, Pine Creek, Lower Powder River, the Upper Snake River Canyon, Rye Valley, Lower Burnt River Valley, Mormon Basin, the Malheur and Clarks Creek districts, Sutton Creek and others. Of particular interest are important details on numerous gold mines and prospects in these mining districts, including their locations, histories, geology and other important information, as well as information on silver, copper and fire opal deposits. **8.5" X 11", 250 ppgs. Retail Price: $24.99**

Mining in the Cascades Range of Oregon - Originally published in 1938, this important publication on Oregon Mining has not been available for over seventy five years. Included in this volume are rare insights into the gold mines and other types of metal mines in the Cascades Mountain Range of Oregon. Some of the important mining areas covered include the famous Bohemia Mining District, the North Santiam Mining District, Quartzville Mining District, Blue River Mining District, Fall Creek Mining District, Oakridge District, Zinc District, Buzzard-Al Sarena District, Grand Cove, Climax District and Barron Mining District. Of particular interest are important details on over 100 mines and prospects in these mining districts, including their locations, histories, geology and other important information. **8.5" X 11", 170 ppgs. Retail Price: $14.99**

Beach Gold Placers of the Oregon Coast - Originally published in 1934, this important publication on Oregon Mining has not been available for over 80 years. Included in this volume are rare insights into the beach gold deposits of the State of Oregon, including their locations, occurance, composition and geology. Of particular interest is information on placer platinum in Oregon's rich beach deposits. Also included are the locations and other information on some famous Oregon beach mines, including the Pioneer, Eagle, Chickamin, Iowa and beach placer mines north of the mouth of the Rogue River. **8.5" X 11", 60 ppgs. Retail Price: $8.99**

Mineralogical Composition of the Sands of the Oregon Coast: From Coos Bay to the Columbia - Published in 1945, he text features hard to find information on the composition of the gold bearing black sands of the South West Oregon Coast, offering a unique insight to prospectors in search of Oregon's legendary beach gold. 104 ppgs, $9.99

Manganese Mining in Oregon - First released in 1942 and now out of print, this special reprint edition of "Manganese in Oregon" was originally published by the Oregon Department of Geology and Mineral Industries. The text features hard to find information on the mining of Manganese in Oregon, including details and maps of Oregon manganese mines and prospects. 108 ppgs, 9.99

Medford Oregon As A Mining Center - Written in 1912, this hard to find publication includes valuable insights into the mining history of South West Oregon. This small book contains interesting information on the gold, copper and mining industry in Southern Oregon as it existed just prior to World War One, shedding light on some of the important mines in the area. Included are rare photographs and vintage advertising of the day. 80 ppgs, 9.99

Mineral Resources of Curry County Oregon - First released in 1977 and now out of print, this special reprint edition of "Geology, Mineral Resources and Rock Materials of Curry County, Oregon" was originally published in cooperation of Curry County, Oregon and the Oregon Department of Geology and Mineral Industries. The text features hard to find information on not only the mining of gold and other metals in Curry County, but also aggregate mining in the area. 102 ppgs, 11.99

Origin of the Gold Bearing Black Sands of the Coast of South West Oregon - First released in 1943 and now out of print, this special reprint edition of "The Origin of the Black Sands of the South West Oregon Coast" was originally published by the Oregon Department of Geology and Mineral Industries. The text features hard to find information on the origin of the gold bearing black sands of the South West Oregon Coast, offering a unique insight to prospectors in search of Oregon's legendary beach gold. 52 ppgs, 8.99

South West Oregon Mining - Leading mining historian Kerby Jackson introduces us to six classic small mining publications on the Gold Mining Industry in Southern Oregon. This small book consists of a compilation of USGS J.S. Diller's "Mines of the Riddles Quadrangle", "The Rogue River Valley Coal Fields" and "Mineral Resources of the Grants Pass Quadrangle", the Grants Pass Commercial Club's rare publication "Mining in Josephine County, Oregon" and the USGS publication "The Distribution of Placer Gold in the Sixes River, South West Oregon". Also included is F.W. Libbey's legendary article on the Southern Oregon Mining Industry, "Lest We Forget", which appeared in the publication of the Oregon State Department of Geology and Mineral Industries in the early 1960's. This compilation offers a unique perspective on mining in South West Oregon and includes considerable information on mines in Josephine, Jackson and Coos Counties. 142 ppgs, 14.99

Geology and Mineral Resources of the Gasquet Quadrangle of California-Oregon - First published in 1953, it has been unavailable for over a century and sheds important light on the geological features and mineral resources of this portion of Northern California and Southern Oregon. 80 ppgs, 9.99

Idaho Mining Books

Gold in Idaho - Unavailable since the 1940's, this publication was originally compiled by the Idaho Bureau of Mines and includes details on gold mining in Idaho. Included is not only raw data on gold production in Idaho, but also valuable insight into where gold may be found in Idaho, as well as practical information on the gold bearing rocks and other geological features that will assist those looking for placer and lode gold in the State of Idaho. This volume also includes thirteen gold maps that greatly enhance the practical usability of the information contained in this small book detailing where to find gold in Idaho. **8.5" X 11", 72 ppgs. Retail Price: $9.99**

Geology of the Couer D'Alene Mining District of Idaho - Unavailable since 1961, this publication was originally compiled by the Idaho Bureau of Mines and Geology and includes details on the mining of gold, silver and other minerals in the famous Coeur D'Alene Mining District in Northern Idaho. Included are details on the early history of the Coeur D'Alene Mining District, local tectonic settings, ore deposit features, information on the mineral belts of the Osburn Fault, as well as detailed information on the famous Bunker Hill Mine, the Dayrock Mine, Galena Mine, Lucky Friday Mine and the infamous Sunshine Mine. This volume also includes sixteen hard to find maps. **8.5" X 11", 70 ppgs. Retail Price: $9.99**

The Gold Camps and Silver Cities of Idaho - Originally published in 1963, this important publication on Idaho Mining has not been available for nearly fifty years. Included are rare insights into the history of Idaho's Gold Rush, as well as the mad craze for silver in the Idaho Panhandle. Documented in fine detail are the early mining excitements at Boise Basin, at South Boise, in the Owyhees, at Deadwood, Long Valley, Stanley Basin and Robinson Bar, at Atlanta, on the famous Boise River, Volcano, Little Smokey, Banner, Boise Ridge, Hailey, Leesburg, Lemhi, Pearl, at South Mountain, Shoup and Ulysses, Yellow Jacket and Loon Creek. The story follows with the appearance of Chinese miners at the new mining camps on the Snake River, Black Pine, Yankee Fork, Bay Horse, Clayton, Heath, Seven Devils, Gibbonsville, Vienna and Sawtooth City. Also included are special sections on the Idaho Lead and Silver mines of the late 1800's, as well as the mining discoveries of the early 1900's that paved the way for Idaho's modern mining and mineral industry. Lavishly illustrated with rare historic photos, this volume provides a one of a kind documentary into Idaho's mining history that is sure to be enjoyed by not only modern miners and prospectors who still scour the hills in search of nature's treasures, but also those enjoy history and tromping through overgrown ghost towns and long abandoned mining camps. **8.5" X 11", 186 ppgs. Retail Price: $14.99**

Ore Deposits and Mining in North Western Custer County Idaho - Unavailable since 1913, this important publication was originally published by the Us Department of the Interior and has been unavailable for a century. Included are fine details on the geology, geography, gold placers and gold and silver bearing quartz veins of the mining region of North West Custer County, Idaho. Of particular interest is a rare look at the mines and prospects of the region, including those such as the Ramshorn Mine, SkyLark, Riverview, Excelsior, Beardsley, Pacific, Hoosier, Silver Brick, Forest Rose and dozens of others in the Bay Horse Mining District. Also covered are the mines of the Yankee Fork District such as the Lucky Boy, Badger, Black, Enterprise, Charles Dickens, Morrison, Golden Sunbeam, Montana, Golden Gate and others, as well as those in the Loon Mining District. **8.5" X 11", 126 ppgs. Retail Price: $12.99**

Gold Rush To Idaho - Unavailable since 1963, this important publication was originally published by the Idaho Bureau of Mines and has been unavailable for 50 years. "Gold Rush To Idaho" revisits the earliest years of the discovery of gold in Idaho Territory and introduces us to the conditions that the pioneer gold seekers met when they blazed a trail through the wilderness of Idaho's mountains and discovered the precious yellow metal at Oro Fino and Pierce. Subsequent rushes followed at places like Elk City, Newsome, Clearwater Station, Florence, Warrens and elsewhere. Of particular interest is a rare look at the hardships that the first miners in Idaho met with during their day to day existences and their attempts to bring law and order to their mining camps. **8.5" X 11", 88 ppgs. Retail Price: $9.99**

The Geology and Mines of Northern Idaho and North Western Montana - Unavailable since 1909, this important publication was originally published by the Us Department of the Interior and has been unavailable for a century. Included are fine details on the geology and geography of the mining regions of Northern Idaho and North Western Montana. Of particular interest is a rare look at the mines and prospects of the region, including those in the Pine Creek Mining District, Lake Pend Oreille district, Troy Mining District, Sylvanite District, Cabinet Mining District, Prospect Mining District and the Missoula Valley. Some of the mines featured include the Iron Mountain, Silver Butte, Snowshoe, Grouse Mountain Mine and others. **8.5" X 11", 142 ppgs. Retail Price: $12.99**

Mining in the Alturas Quadrangle of Blaine County Idaho - Unavailable since 1922, this important publication was originally published by the Idaho Bureau of Mines and has been unavailable for ninety years. Topics include the geology, rock formations and the formation of ore deposits in this important mining area of Idaho. Of particular focus is information on the local geology, quartz veins and ore deposits of this portion of Idaho. Included are hard to find details, including the descriptions and locations of numerous gold and silver mines in the area including the Silver King, Pilgrim, Columbia, Lone Jack, Sunbeam, Pride of the West, Lucky Boy, Scotia, Atlanta, Beaver-Bidwell and others mines and prospects. **8.5" X 11", 56 ppgs. Retail Price: $8.99**

Mining in Lemhi County Idaho - Originally published in 1913, this important book on Idaho Mining has not been available to miners for over a century. Included are rare insights into hundreds of gold, silver, copper and other mines in this famous Idaho mining area. Details include the locations, geology, history, production and other facts of the mines of this region, not only gold and silver hardrock mines, but also gold placer mines, lead-silver deposits, copper mines, cobalt-nickel deposits, tungsten and tin mines . It is lavishly illustrated with hard to find photos of the period and rare mining maps. Some of the vicinities featured include the Nicholia Mining District, Spring Mountain District, Texas District, Blue Wing District, Junction District, McDevitt District, Pratt Creek, Eldorado District, Kirtley Creek, Carmen Creek, Gibbonsville, Indian Creek, Mineral Hill District, Mackinaw, Eureka District, Blackbird District, YellowJacket District, Gravel Range District, Junction District, Parker Mountain and other mining districts. **8.5" X 11", 226 ppgs. Retail Price: $19.99**

Mining in Shoshone County Idaho - First published in 1923, it has been unavailable for over a century and sheds important light on the mining history of Shoshone County, Idaho. Some of the topics include the history of mining in Shoshone County, a look at the local geology and ore characteristics of lead-silver deposits, zinc deposits, copper, antimony, gold and other minerals. Also included are insights into the history, production, characteristics and locations of numerous mines in the area. 198 ppgs, 15.99

Utah Mining Books

Fluorite in Utah - Unavailable since 1954, this publication was originally compiled by the USGS, State of Utah and U.S. Atomic Energy Commission and details the mining of fluorspar, also known as fluorite in the State of Utah. Included are details on the geology and history of fluorspar (fluorite) mining in Utah, including details on where this unique gem mineral may be found in the State of Utah. 8.5" X 11", 60 ppgs. Retail Price: $8.99

The Gold Hill Mining District of Utah - First published in 1935, it has been unavailable since those days and sheds important light on the mines, history and geology of Utah's Gold Hill Mining District. Included are rare insights into this important mining area, including the locations, histories and details of numerous mines. This volume is well illustrated with geological diagrams, as well as hard to find maps of some of the most important mines in this district. 202 ppgs., 19.99

The Mines, Miners and Minerals of Utah - First published in 1896, it has been unavailable since those days and sheds important light on the early mines and miners of Pioneer Utah, as well as the minerals which they won from the earth by laborious hard physical labor and sheer determination. Included are rare insights into the early mining history of Utah, as well details on hundreds of gold, silver and copper mines. 376 ppgs., 24.99

California Mining Books

The Tertiary Gravels of the Sierra Nevada of California - Mining historian Kerby Jackson introduces us to a classic mining work by Waldemar Lindgren in this important re-issue of The Tertiary Gravels of the Sierra Nevada of California. Unavailable since 1911, this publication includes details on the gold bearing ancient river channels of the famous Sierra Nevada region of California. 8.5" X 11", 282 ppgs. Retail Price: $19.99

The Mother Lode Mining Region of California - Unavailable since 1900, this publication includes details on the gold mines of California's famous Mother Lode gold mining area. Included are details on the geology, history and important gold mines of the region, as well as insights into historic mining methods, mine timbering, mining machinery, mining bell signals and other details on how these mines operated. Also included are insights into the gold mines of the California Mother Lode that were in operation during the first sixty years of California's mining history. **8.5" X 11", 176 ppgs. Retail Price: $14.99**

Lode Gold of the Klamath Mountains of Northern California and South West Oregon - Unavailable since 1971, this publication was originally compiled by Preston E. Hotz and includes details on the lode mining districts of Oregon and California's Klamath Mountains. Included are details on the geology, history and important lode mines of the French Gulch, Deadwood, Whiskeytown, Shasta, Redding, Muletown, South Fork, Old Diggings, Dog Creek (Delta), Bully Choop (Indian Creek), Harrison Gulch, Hayfork, Minersville, Trinity Center, Canyon Creek, East Fork, New River, Denny, Liberty (Black Bear), Cecilville, Callahan, Yreka, Fort Jones and Happy Camp mining districts in California, as well as the Ashland, Rogue River, Applegate, Illinois River, Takilma, Greenback, Galice, Silver Peak, Myrtle Creek and Mule Creek districts of South Western Oregon. Also included are insights into the mineralization and other characteristics of this important mining region. **8.5" X 11", 100 ppgs. Retail Price: $10.99**

Mines and Mineral Resources of Shasta County, Siskiyou County, Trinity County: California - Unavailable since 1915, this publication was originally compiled by the California State Mining Bureau and includes details on the gold mines of this area of Northern California. Also included are insights into the mineralization and other characteristics of this important mining region, as well as the location of historic gold mines. **8.5" X 11", 204 ppgs. Retail Price: $19.99**

Geology of the Yreka Quadrangle, Siskiyou County, California - Unavailable since 1977, this publication was originally compiled by Preston E. Hotz and includes details on the geology of the Yreka Quadrangle of Siskiyou County, California. Also included are insights into the mineralization and other characteristics of this important mining region. **8.5" X 11", 78 ppgs. Retail Price: $7.99**

Mines of San Diego and Imperial Counties, California - Originally published in 1914, this important publication on California Mining has not been available for a century. This publication includes important information on the early gold mines of San Diego and Imperial County, which were some of the first gold fields mined in California by early Spanish and Mexican miners before the 49ers came on the scene. Included are not only details on early mining methods in the area, production statistics and geological information, but also the location of the early gold mines that helped make California "The Golden State". Also included are details on the mining of other minerals such as silver, lead, zinc, manganese, tungsten, vanadium, asbestos, barite, borax, cement, clay, dolomite, fluospar, gem stones, graphite, marble, salines, petroleum, stronium, talc and others. **8.5" X 11", 116 ppgs. Retail Price: $12.99**

Mines of Sierra County, California - Unavailable since 1920, this publication was originally compiled by the California State Mining Bureau and includes details on the gold mines of Sierra County, California. Also included are insights into the mineralization and other characteristics of this important mining region, as well as the location of historic gold mines. **8.5" X 11", 156 ppgs. Retail Price: $19.99**

Mines of Plumas County, California - Unavailable since 1918, this publication was originally compiled by the California State Mining Bureau and includes details on the gold mines of Plumas County, California. Also included are insights into the mineralization and other characteristics of this important mining region, as well as the location of historic gold mines. **8.5" X 11", 200 ppgs. Retail Price: $19.99**

Mines of El Dorado, Placer, Sacramento and Yuba Counties, California - Originally published in 1917, this important publication on California Mining has not been available for nearly a century. This publication includes important information on the early gold mines of El Dorado County, Placer County, Sacramento County and Yuba County, which were some of the first gold fields mined by the Forty-Niners during the California Gold Rush. Included are not only details on early mining methods in the area, production statistics and geological information, but also the location of the early gold mines that helped make California "The Golden State". Also included are insights into the early mining of chrome, copper and other minerals in this important mining area. **8.5" X 11", 204 ppgs. Retail Price: $19.99**

Mines of Los Angeles, Orange and Riverside Counties, California - Originally published in 1917, this important publication on California Mining has not been available for nearly a century. This publication includes important information on the early gold mines of Los Angeles County, Orange County and Riverside County, which were some of the first gold fields mined in California by early Spanish and Mexican miners before the 49ers came on the scene. Included are not only details on early mining methods in the area, production statistics and geological information, but also the location of the early gold mines that helped make California "The Golden State". **8.5" X 11", 146 ppgs. Retail Price: $12.99**

Mines of San Bernadino and Tulare Counties, California - Originally published in 1917, this important publication on California Mining has not been available for nearly a century. This publication includes important information on the early gold mines of San Bernadino and Tulare County, which were some of the first gold fields mined in California by early Spanish and Mexican miners before the 49ers came on the scene. Included are not only details on early mining methods in the area, production statistics and geological information, but also the location of the early gold mines that helped make California "The Golden State". Also included are details on the mining of other minerals such as copper, iron, lead, zinc, manganese, tungsten, vanadium, asbestos, barite, borax, cement, clay, dolomite, fluospar, gem stones, graphite, marble, salines, petroleum, stronium, talc and others. **8.5" X 11", 200 ppgs. Retail Price: $19.99**

Chromite Mining in The Klamath Mountains of California and Oregon - Unavailable since 1919, this publication was originally compiled by J.S. Diller of the United States Department of Geological Survey and includes details on the chromite mines of this area of Northern California and Southern Oregon. Also included are insights into the mineralization and other characteristics of this important mining region, as well as the location of historic mines. Also included are insights into chromite mining in Eastern Oregon and Montana. **8.5" X 11", 98 ppgs. Retail Price: $9.99**

Mines and Mining in Amador, Calaveras and Tuolumne Counties, California - Unavailable since 1915, this publication was originally compiled by William Tucker and includes details on the mines and mineral resources of this important California mining area. Included are details on the geology, history and important gold mines of the region, as well as insights into other local mineral resources such as asbestos, clay, copper, talc, limestone and others. Also included are insights into the mineralization and other characteristics of this important portion of California's Mother Lode mining region. **8.5" X 11", 198 ppgs. Retail Price: $14.99**

The Cerro Gordo Mining District of Inyo County California - Unavailable since 1963, this publication was originally compiled by the United States Department of Interior. Included are insights into the mineralization and other characteristics of this important mining region of Southern California. Topics include the mining of gold and silver in this important mining district in Inyo County, California, including details on the history, production and locations of the Cerro Gordo Mine, the Morning Star Mine, Estelle Tunnel, Charles Lease Tunnel, Ignacio, Hart, Crosscut Tunnel, Sunset, Upper Newtown, Newtown, Ella, Perseverance, Newsboy, Belmont and other silver and gold mines in the Cerro Gordo Mining District. This volume also includes important insights into the fossil record, geologic formations, faults and other aspects of economic geology in this California mining district. **8.5" X 11", 104 ppgs. Retail Price: $10.99**

Mining in Butte, Lassen, Modoc, Sutter and Tehama Counties of California - Unavailable since 1917, this publication was originally compiled by the United States Department of Interior. Included are insights into the mineralization and other characteristics of this important mining region of California. Topics include the mining of asbestos, chromite, gold, diamonds and manganese in Butte County, the mining of gold and copper in the Hayden Hill and Diamond Mountain mining districts of Lassen County, the mining of coal, salt, copper and gold in the High Grade and Winters mining districts of Modoc County, gold mining in Sutter County and the mining of gold, chromite, manganese and copper in Tehama County. This volume also includes the production records and locations of numerous mines in this important mining region. **8.5" X 11", 114 ppgs. Retail Price: $11.99**

Mines of Trinity County California - Originally published in 1965, this important publication on California Mining has not been available for nearly fifty years. This publication includes important information on mines and mining in Trinity County, California, as well insights into the mineralization and geology of this important mining area in Northern California. Included are extensive details on hardrock and placer gold mines and prospects, including charts showing the locations of these historic mines.. **8.5" X 11", 144 ppgs. Retail Price: $12.99**

Mines of Kern County California - Originally published in 1962, this important publication on California Mining has not been available for nearly fifty years. This publication includes important information on mines and mining in Kern County, California, as well insights into the mineralization and geology of this important mining area in California. Included are extensive details on hardrock and placer gold mines and prospects, including charts showing the locations of these historic mines. **8.5" X 11", 398 ppgs. Retail Price: $24.99**

Mines of Calaveras County California - Originally published in 1962, this important publication on California Mining has not been available for nearly fifty years. This publication includes important information on mines and mining in Calaveras County, California, as well insights into the mineralization and geology of this important mining area in Northern California. Included are extensive details on hardrock and placer gold mines and prospects, including charts showing the locations of these historic mines. **8.5" X 11", 236 ppgs. Retail Price: $19.99**

Lode Gold Mining in Grass Valley California - Unavailable since 1940, this publication was originally compiled by the United States Department of Interior. Included are insights into the gold mineralization and other characteristics of this important mining region of Nevada County, California. This volume also includes important insights into the geologic formations, faults and other aspects of economic geology in this California mining district. Of particular interest are the fine details on many hardrock gold mines in the area, including their locations, histories, development and mineralization. Some of the mines featured include the Gold Hill Mine, Massachusetts Hill, Boundary, Peabody, Golden Center, North Star, Omaha, Lone Jack, Homeward Bound, Hartery, Wisconsin, Allison Ranch, Phoenix, Kate Hayes, W.Y.O.D., Empire, Rich Hill, Daisy Hill, Orleans, Sultana, Centennial, Conlin, Ben Franklin, Crown Point and many others. **8.5" X 11", 148 ppgs. Retail Price: $12.99**

Lode Mining in the Alleghany District of Sierra County California - Unavailable since 1913, this publication was originally compiled by the United States Department of Interior. Included are insights into the mineralization and other characteristics of this important mining region of Sierra County. Included are details on the history, production and locations of numerous hardrock gold mines in this famous California area, including the Tightner Mine, Minnie D., Osceola, Eldorado, Twenty One, Sherman, Kenton, Oriental, Rainbow, Plumbago, Irelan, Gold Canyon, North Fork, Federal, Kate Hardy and others. This volume also includes important insights into the fossil record, geologic formations, faults and other aspects of economic geology in this California mining district. **8.5" X 11", 48 ppgs. Retail Price: $7.99**

Six Months In The Gold Mines During The California Gold Rush - Unavailable since 1850, this important work is a first hand account of one "49'ers" personal experience during the great California Gold Rush, shedding important light on one of the most exciting periods in the history of not only California, but also the world. Compiled from journals written between 1847 and 1849 by E. Gould Buffum, a native of New York, "Six Months In The Gold Mines During The California Gold Rush" offers a rare look into the day to day lives of the people who came to California to work in her gold mines when the state was still a great frontier. **8.5" X 11", 290 ppgs. Retail Price: $19.99**

<u>Quartz Mines of the Grass Valley Mining District of California</u> - Unavailable since 1867, this important publication has not been available since those days. This rare publication offers a short dissertation on the early hardrock mines in this important mining district in the California Mother Lode region between the 1850's and 1860's. Also included are hard to find details on the mineralization and locations of these mines, as well as how they were operated in those day. **8.5" X 11", 44 ppgs. Retail Price: $8.99**

<u>Gold Rush on the Feather River</u> - First published in 1924, this short publication by G.C. Mansfield sheds important light on the early history of gold mining on the Feather River. Included are rare insights into the first decade of gold mining and the early mining camps of the Feather River during the 1850's. **64 ppgs., 9.99**

<u>The Bodie Mining District of California</u> - First published in 1986, it has been unavailable since those days and sheds important light on this famous mining area. Included are the history, characteristics and locations of numerous old mines around the ghost town of Bodie.
64 ppgs, 8.99

<u>Geology and Mineral Resources of the Gasquet Quadrangle of California-Oregon</u> - First published in 1953, it has been unavailable for over a century and sheds important light on the geological features and mineral resources of this portion of Northern California and Southern Oregon.
80 ppgs, 9.99

Alaska Mining Books

<u>Ore Deposits of the Willow Creek Mining District, Alaska</u> - Unavailable since 1954, this hard to find publication includes valuable insights into the Willow Creek Mining District near Hatcher Pass in Alaska. The publication includes insights into the history, geology and locations of the well known mines in the area, including the Gold Cord, Independence, Fern, Mabel, Lonesome, Snowbird, Schroff-O'Neil, High Grade, Marion Twin, Thorpe, Webfoot, Kelly-Willow, Lane, Holland and others. **8.5" X 11", 96 ppgs. Retail Price: $9.99**

<u>The Juneau Gold Belt of Alaska</u> - Unavailable since 1906, this hard to find publication includes valuable insights into the gold mines around Juneau, Alaska. The publication includes important details into the history, geology and locations of the well known gold mines and prospects in the area, including those around Windham Bay, Holkham Bay, Port Snettisham, on Grindstone and Rhine Creeks, Gold Creek, Douglas Island, Salmon Creek, Lemon Creek, Nugget Creek, from the Mendenhall River to Berners Bay, McGinnis Creek, Montana Creek, Peterson Creek, Windfall Creek, the Eagle River, Yankee Basin, Yankee Curve, Kowee Creek and elsewhere. Not only are gold placer mines included, but also hardrock gold mines. **8.5" X 11", 224 ppgs. Retail Price: $19.99**

<u>Mining in the Jumbo Basin of Alaska</u> - Unavailable since 1953, this hard to find publication includes valuable insights into the mines and geology of the Jumbo Basin. The publication includes important details into the history, geology and locations of the well known gold mines and prospects in the famous Jumbo Basin Mining Region of Alaska.
72 ppgs, 9.99

<u>The Rampart Placer Gold Region of Alaska</u> - Unavailable since 1906, this hard to find publication includes valuable insights into the placer gold mines of the Rampart Mining Region. The publication includes important details into the history, geology and locations of the well known gold mines and prospects in the famous Rampart Mining Region of Alaska.
78 ppgs, 10.99

Arizona Mining Books

<u>Mines and Mining in Northern Yuma County Arizona</u> - Originally published in 1911, this important publication on Arizona Mining has not been available for over a hundred years. Included are rare insights into the gold, silver, copper and quicksilver mines of Yuma County, Arizona together with hard to find maps and photographs. Some of the mines and mining districts featured include the Planet Copper Mine, Mineral Hill, the Clara Consolidated Mine, Viati Mine, Copper Basin prospect, Bowman Mine, Quartz King, Billy Mack, Carnation, the Wardwell and Osbourne, Valensuella Copper, the Mariquita, Colonial Mine, the French American, the New York-Plomosa, Guadalupe, Lead Camp, Mudersbach Copper Camp, Yellow Bird, the Arizona Northern (Salome Strike), Bonanza (Harqua Hala), Golden Eagle, Hercules, Socorro and others. **8.5" X 11", 144 ppgs. Retail Price: $11.99**

<u>The Aravaipa and Stanley Mining Districts of Graham County Arizona</u> - Originally published in 1925, this important publication on Arizona Mining has not been available for nearly ninety years. Included are rare insights into the gold and silver mines of these two important mining districts, together with hard to find maps. **8.5" X 11", 140 ppgs. Retail Price: $11.99**

Gold in the Gold Basin and Lost Basin Mining Districts of Mohave County, Arizona - This volume contains rare insights into the geology and gold mineralization of the Gold Basin and Lost Basin Mining Districts of Mohave County, Arizona that will be of benefit to miners and prospectors. Also included is a significant body of information on the gold mines and prospects of this portion of Arizona. This volume is lavishly illustrated with rare photos and mining maps. **8.5" X 11", 188 ppgs. Retail Price: $19.99**

Mines of the Jerome and Bradshaw Mountains of Arizona - This important publication on Arizona Mining has not been available for ninety years. This volume contains rare insights into the geology and ore deposits of the Jerome and Bradshaw Mountains of Arizona that will be of benefit to miners and prospectors who work those areas. Included is a significant body of information on the mines and prospects of the Verde, Black Hills, Cherry Creek, Prescott, Walker, Groom Creek, Hassayampa, Bigbug, Turkey Creek, Agua Fria, Black Canyon, Peck, Tiger, Pine Grove, Bradshaw, Tintop, Humbug and Castle Creek Mining Districts. This volume is lavishly illustrated with rare photos and mining maps. **8.5" X 11", 218 ppgs. Retail Price: $19.99**

The Ajo Mining District of Pima County Arizona - This important publication on Arizona Mining has not been available for nearly seventy years. This volume contains rare insights into the geology and mineralization of the Ajo Mining District in Pima County, Arizona and in particular the famous New Cornelia Mine. **8.5" X 11", 126 ppgs. Retail Price: $11.99**

Mining in the Santa Rita and Patagonia Mountains of Arizona - Originally published in 1915, this important publication on Arizona Mining has not been available for nearly a century. Included are rare insights into hundreds of gold, silver, copper and other mines in this famous Arizona mining area. Details include the locations, geology, history, production and other facts of the mines of this region. **8.5" X 11", 394 ppgs. Retail Price: $24.99**

Mining in the Bisbee Quadrangle of Arizona - Originally published in 1906, this important publication on Arizona Mining has not been available for nearly a century. Included are rare insights into hundreds of gold, silver, copper and other mines in this famous Arizona mining area. Details include the locations, geology, history, production and other facts of the mines of this important mining region. **8.5" X 11", 188 ppgs. Retail Price: $14.99**

Placer Gold Mining in Arizona - Unavailable since 1922, this hard to find publication includes valuable insights into the placer gold mines of the Arizona. Originally released as "Placer Gold of Arizona", despite its small size, this publication includes important details into the history, geology and locations of the well known placer gold mines and prospects in the State of Arizona. **48 ppgs, 8.99**

Gold and Copper Mining near Payson, Arizona - Written in 1915, this hard to find publication includes valuable insights into the gold and copper mining industry of Arizona. Highlighted here are the gold and copper mines near Payson, Arizona. **68 ppgs, 8.99**

Lode Gold Mining in Arizona - Unavailable since 1934, this hard to find publication, originally released as "Arizona Lode Gold Mines and Gold Mining" includes valuable insights into the gold mining industry of Arizona. Included are valuable insights into over 150 hardrock gold mines in over 30 different mining districts in Arizona. **278 ppgs, 21.99**

Mining in the Dragoon Quadrangle of Cochise County, Arizona - Unavailable since 1964, this hard to find publication includes valuable insights into the mines of the Dragoon Quadrangle Mining Region. The publication includes important details into the history, geology and locations of the well known mines and prospects in this famous mining region of Arizona. **224 ppgs., 19.99**

Directory of Operating Mines in Arizona in 1915 - Unavailable since 1916, this hard to find publication includes valuable insights into the mines of Arizona. This small publication includes a complete list of the mines that were operating in the State of Arizona during 1915 and includes details such as general location, owners and some basic facts about each mining operation. **52 ppgs. 8.99**

Arizona Ore Deposits - Unavailable since 1938, this hard to find publication includes valuable insights into some ore deposits of Arizona. Included are valuable insights into the formation and characteristics of valuable ore deposits in the Jerome, Miami, Inspiration, Clifton, Morenci, Ray, Ajo, Eureka, Tombstone and Magma mining districts. Included are details into some of the major gold, silver and copper mines of these important Arizona mining areas. **160 ppgs, 14.99**

Montana Mining Books

A History of Butte Montana: The World's Greatest Mining Camp - First published in 1900 by H.C. Freeman, this important publication sheds a bright light on one of the most important mining areas in the history of The West. Together with his insights, as well as rare photographs of the periods, Harry Freeman describes Butte and its vicinity from its early beginnings, right up to its flush years when copper flowed from its mines like a river. At the time of publication, Butte, Montana was known worldwide as "The Richest Mining Spot On Earth" and produced not only vast amounts of copper, but also silver, gold and other metals from its mines. Freeman illustrates, with great detail, the most important mines in the vicinity of Butte, providing rare details on their owners, their history and most importantly, how the mines operated and how their treasures were extracted. Of particular interest are the dozens of rare photographs that depict mines such as the famous Anaconda, the Silver Bow, the Smoke House, Moose, Paulin, Buffalo, Little Minah, the Mountain Consolidated, West Greyrock, Cora, the Green Mountain, Diamond, Bell, Parnell, the Neversweat, Nipper, Original and many others. **8.5" X 11", 142 ppgs. Retail Price: $12.99**

The Butte Mining District of Montana - This important publication on Montana Mining has not been available for over a century. Included are rare insights into the gold, copper and silver mines of Butte, Montana together with hard to find maps and photographs. Some of the topics include the early history of gold, silver and copper mining in the Butte area, insight into the geology of its mining areas, the local distribution of gold, silver and copper ores, as well their composition and how to identify them. Also included are detailed facts about the mines in the Butte Mining District, including the famous Anaconda Mine, Gagnon, Parrot, Blue Vein, Moscow, Poulin, Stella, Buffalo, Green Mountain, Wake Up Jim, the Diamond-Bell Group, Mountain Consolidated, East Greyrock, West Greyrock, Snowball, Corra, Speculator, Adirondack, Miners Union, the Jessie-Edith May Group, Otisco, Iduna, Colorado, Lizzie, Cambers, Anderson, Hesperus, Preferencia and dozens of others. **8.5" X 11", 298 ppgs. Retail Price: $24.99**

Mines of the Helena Mining Region of Montana - This important publication on Montana Mining has not been available for over a century. Included are rare insights into the gold, copper and silver mines of the vicinity of Helena, Montana, including the Marysville Mining District, Elliston Mining District, Rimini Mining District, Helena Mining District, Clancy Mining District, Wickes Mining District, Boulder and Basin Mining Districts and the Elkhorn Mining District. Some of the topics include the early history of gold, silver and copper mining in the Helena area, insight into the geology of its mining areas, the local distribution of gold, silver and copper ores, as well their composition and how to identify them. Also included are detailed facts, history, geology and locations of over one hundred gold, silver and copper mines in the area . **8.5" X 11", 162 ppgs, Retail Price: $14.99**

Mines and Geology of the Garnet Range of Montana - This important publication on Montana Mining has not been available for over a century. Included are rare insights into the gold, copper and silver mines of the vicinity of this important mining area of Montana. Some of the topics include the early history of gold, silver and copper mining in the Garnet Mountains, insight into the geology of its mining areas, the local distribution of gold, silver and copper ores, as well their composition and how to identify them. Also included are detailed facts, history, geology and locations of numerous gold, silver and copper mines in the area . **8.5" X 11", 100 ppgs, Retail Price: $11.99**

Mines and Geology of the Philipsburg Quadrangle of Montana - This important publication on Montana Mining has not been available for over a century. Included are rare insights into the gold, copper and silver mines of the vicinity of this important mining area of Montana. Some of the topics include the early history of gold, silver and copper mining in the Philipsburg Quadrangle, insight into the geology of its mining areas, the local distribution of gold, silver and copper ores, as well their composition and how to identify them. Also included are detailed facts, history, geology and locations of over one hundred gold, silver and copper mines in the area **8.5" X 11", 290 ppgs, Retail Price: $24.99**

Geology of the Marysville Mining District of Montana - Included are rare insights into the mining geology of the Marysville Mining District. Some of the topics include the early history of gold, silver and copper mining in the area, insight into the geology of its mining areas, the local distribution of gold, silver and copper ores, as well their composition and how to identify them. Also included are detailed facts, history, geology and locations of gold, silver and copper mines in the area **8.5" X 11", 198 ppgs, Retail Price: $19.99**

The Geology and Mines of Northern Idaho and North Western Montana- See listing under Idaho.

The History of Gold Dredging in Montana - Unavailable since 1916, this important publication was originally published by the Us Bureau of Mines and has been unavailable for a century. A century and more ago, giant dredging machines dug in Montana's rivers and creeks in search of illusive golden riches. First appearing in California in the 1850's, gold dredges finally reached their peak of development in Siberia and New Zealand before becoming popular again in the United States. This book offers a unique historical perspective on the gold dredges that once operated in Montana. This book on Montana mining history is lavishly illustrated with dozens of rare historic photos gold dredges that once operated in Montana, as well as hard to locate plans on how these dredges were designed. 120 ppgs., 11.99

Nevada Mining Books

The Bull Frog Mining District of Nevada - Unavailable since 1910, this publication was originally compiled by the United States Department of Interior. This volume also includes important insights into the geologic formations, faults and other aspects of economic geology in this Nevada mining district. Of particular interest are the fine details on many mines in the area, including their locations, histories, development and mineralization. Some of the mines featured include the National Bank Mine, Providence, Gibraltor, Tramps, Denver, Original Bullfrog, Gold Bar, Mayflower, Homestake-King and other mines and prospects. **8.5" X 11", 152 ppgs, Retail Price: $14.99**

History of the Comstock Lode - Unavailable since 1876, this publication was originally released by John Wiley & Sons. This volume also includes important insights into the famous Comstock Lode of Nevada that represented the first major silver discovery in the United States. During its spectacular run, the Comstock produced over 192 million ounces of silver and 8.2 million ounces of gold. Not only did the Comstock result in one of the largest mining rushes in history and yield immense fortunes for its owners, but it made important contributions to the development of the State of Nevada, as well as neighboring California. Included here are important details on not only the early development and history of the Comstock, but also rare early insight into its mines, ore and its geology.**8.5" X 11", 244 ppgs, Retail Price: $19.99**

The Pioche Mining District of Nevada - First published in 1932, it has been unavailable for over a century and sheds important light on the mining history of Nevada. Some of the topics include the history of mining in this district, as well as the characteristics of its mineral and ore deposits. Also included are insights into the history, production, characteristics and locations of numerous mines in the area. Some of the mines include the Combined Metals, Pioche, Ely Valley, No. 10, Poorman, Wide Awake, Alps, Prince, Virginia Louise, Half Moon, Abe Lincoln, Fairview, Bristol Silver, National, Vesuvius, Inman, Tempest, Hillside, Jackrabbit, Lucky Star, Fortuna, Mendha, Manhattan, Hamburg, Comet, Lyndon and others. 108 ppgs 10.99

The Yerington Mining District of Nevada - First published in 1932, it has been unavailable for over a century and sheds important light on the mining history of Nevada. Some of the topics include the history of mining in this district, as well as the characteristics of its mineral and ore deposits. Also included are insights into the history, production, characteristics and locations of numerous mines in the area. Some of the mines include the Bluestone, Mason Valley, Malachite, McConnell, Greenwood, Western Nevada, Ludwig, Douglas Hill, Casting Copper, Montana-Yerington, Empire, Jim Beatty, Terry and McFarland, Blue Jay and others. 92 ppgs, 10.99

The Genesis of the Ores of Tonopah Nevada - Unavailable since 1918, this hard to find publication includes valuable insights into the gold mines around Tonopah, Nevada. The publication includes important details into the geology of mines in the Tonopah Mining District of Nevada. 90 ppgs, 10.99

Mining Camps of Elko, Lander and Eureka Counties Nevada - Unavailable since 1910, this hard to find publication includes valuable insights into the mining camps of Elko, Lander and Eureka Counties, Nevada. The publication includes important details into the history of mines and mining in these three Nevada counties. 154 ppgs, 12.99

Ore Deposits of the Bullfrog Quadrangle - Unavailable since 1964 and released as "Geology of Bullfrog Quadrangle and Ore Deposits Related to Bullfrog Hills Caldera, Nye County, Nevada and Inyo County, California". The publication includes important details into the geology of mines in the Bullfrog Quadrangle of Nye County, Nevada and Inyo County, California. 52 ppgs, 9.99

Mining in Eureka County Nevada - Unavailable since 1879, this hard to find publication includes valuable insights into the early mining history off Eureka County, Nevada. The publication includes important details into the early history of the mines of Eureka County, as well as their development, production and how their ores were treated. Also included are details on the 1872 Mining Act, as well as the local rules, regulations and customs of the miners in Eureka County.134 ppgs, 12.99

Colorado Mining Books

Ores of The Leadville Mining District - Unavailable since 1926, this publication was originally compiled by the United States Department of Interior. This volume also includes important insights into the ores and mineralization of the Leadville Mining District in Colorado. Topics include historic ore prospecting methods, local geology, insights into ore veins and stockworks, the local trend and distribution of ore channels, reverse faults, shattered rock above replacement ore bodies, mineral enrichment in oxidized and sulphide zones and more. **8.5" X 11", 66 ppgs, Retail Price: $8.99**

Mining in Colorado - Unavailable since 1926, this publication was originally compiled by the United States Department of Interior. This volume also includes important insights into the mining history of Colorado from its early beginnings in the 1850's right up to the mid 1920's. Not only is Colorado's gold mining heritage included, but also its silver, copper, lead and zinc mining industry. Each mining area is treated separately, detailing the development of Colorado's mines on a county by county basis. **8.5" X 11", 284 ppgs, Retail Price: $19.99**

Gold Mining in Gilpin County Colorado - Unavailable since 1876, this publication was originally compiled by the Register Steam Printing House of Central City, Colorado. A rare glimpse at the gold mining history and early mines of Gilpin County, Colorado from their first discovery in the 1850's up to the "flush years" of the mid 1870's. Of particular interest is the history of the discovery of gold in Gilpin County and details about the men who made those first strikes. Special focus is given to the early gold mines and first mining districts of the area, many of which are not detailed in other books on Colorado's gold mining history. **8.5" X 11", 156 ppgs, Retail Price: $12.99**

Mining in the Gold Brick Mining District of Colorado - Important insights into the history of the Gold Brick Mining District, as well as its local geography and economic geology. Also included are the histories and locations of historic mines in this important Colorado Mining District, including the Cortland, Carter, Raymond, Gold Links, Sacramento, Bassick, Sandy Hook, Chronicle, Grand Prize, Chloride, Granite Mountain, Lucille, Gray Mountain, Hilltop, Maggie Mitchell, Silver Islet, Revenue, Roosevelt, Carbonate King and others. In addition to hardrock mining, are also included are details on gold placer mining in this portion of Colorado. **8.5" X 11", 140 ppgs, Retail Price: $12.99**

Ore Deposits of the London Fault of Colorado - First published in 1941, it has been unavailable since those days and sheds important light on the mines and mineral deposits of the London Fault in Central Colorado's Alma Mining District. This publication sheds important light on the gold veins and lead-silver deposits of the Alma Mining District. Included are geologic details on the London Mine, American Mine, Havigorst Tunnel, Ophir Mine, Mosher Tunnel, London-Butte Mine, Venture Shaft, Hard-To-Beat Mine, Oliver Twist Tunnel, Sacramento Mine, Mudsill Mine, Sherwood Mine, Wagner, Barcoe Tunnel and other mines in this important mining region. 110 ppgs., 10.99

The Mines of Colorado - First published in 1867, it has been unavailable since those days and sheds important light on Colorado's early mining history. Written shortly after the events took place, this publication sheds important light on the Pike's Peak Gold Rush, the discovery of gold on Ralston Creek and Dry Creek in the 1850's, as well as details on the first wave of miners into Colorado and their trials and tribulations as they crossed the Great Plains. Also included are details on early discoveries of lode gold in the mountainous regions of Colorado, details on the early mines hardrock and placer mines, and much more. It is a veritable treasure trove on Colorado's early mining history and will be of great importance to anyone who is interested in the mining of gold or other minerals in Colorado, as well as those interested in the history of the state. 478 ppgs., 29.99

The La Plata Mining District of Colorado - Originally titled "Geology and Ore Deposits in the Vicinity of the La Plata District of Colorado" and first published in 1949, it has been unavailable since those days and sheds important light on the mines and mineral deposits of the La Plata Mining District of Colorado. 214 ppgs., 19.99

Washington Mining Books

The Republic Mining District of Washington - Unavailable since 1910, this important publication was originally published by the Washington Geologic Survey and has been unavailable for a century. Topics include the geology, rock formations and the formation of ore deposits in this important mining area of Washington State. Also included are hard to find details on the geology, history and locations of dozens of mines in the area. Some of the mines featured include the New Republic Mine, Ben Hur, Morning Glory, the South Republic Mine, Quilp, Surprise, Black Tail, Lone Pine, San Poil, Mountain Lion, Tom Thumb, Elcaliph and many others. **8.5" X 11", 94 ppgs, Retail Price: $10.99**

The Myers Creek and Nighthawk Mining Districts of Washington - Unavailable since 1911, this important publication was originally published by the Washington Geologic Survey and has been unavailable for a century. Topics include the geology, rock formations and the formation of ore deposits in these important mining areas of Washington State. Also included are hard to find details on the geology, history and locations of dozens of mines in the area. Some of the mines featured include the Grant Mine, Monterey, Nip and Tuck, Myers Creek, Number Nine, Neutral, Rainbow, Aztec, Crystal Butte, Apex, Butcher Boy, Molson, Mad River, Olentangy, Delate, Kelsey, Golden Chariot, Okanogan, Ohio, Forty-Ninth Parallel, Nighthawk, Favorite, Little Chopaka, Summit, Number One, California, Peerless, Caaba, Prize Group, Ruby, Mountain Sheep, Golden Zone, Rich Bar, Similkameen, Kimberly, Triune, Hiawatha, Trinity, Hornsilver, Maquae, Bellevue, Bullfrog, Palmer Lake, Ivanhoe, Copper World and many others. **8.5" X 11", 136 ppgs, Retail Price: $12.99**

The Blewett Mining District of Washington - Unavailable since 1911, this important publication was originally published by the Washington Geologic Survey and has been unavailable for a century. Topics include the geology, rock formations and the formation of ore deposits in this important mining area of Washington State. Also included are hard to find details on the geology, history and locations of dozens of mines in the area. Some of the mines featured include the Washington Meteor, Alta Vista, Pole Pick, Blinn, North Star, Golden Eagle, Tip Top, Wilder, Golden Guinea, Lucky Queen, Blue Bell, Prospect, Homestake, Lone Rock, Johnson, and others. **8.5" X 11", 134 ppgs, Retail Price: $12.99**

Silver Mining In Washington - Unavailable since 1955, this important publication was originally published by the Washington Geologic Survey. Featured are the hard to find locations and details pertaining to Washington's silver mines. **8.5" X 11", 180 ppgs, Retail Price: $15.99**

The Mines of Snohomish County Washington - Unavailable since 1942, this important publication was originally published by the Washington Geologic Survey and has been unavailable for seventy years. Featured are details on a large number of gold, silver, copper, lead and other metallic mineral mines. Included are the locations of each historic mine, along with information on the commodity produced. **8.5" X 11", 98 ppgs, Retail Price: $10.99**

The Mines of Chelan County Washington - Unavailable since 1943, this important publication was originally published by the Washington Geologic Survey and has been unavailable for seventy years. Featured are details on a large number of gold, silver, copper, lead and other metallic mineral mines. Included are the locations of each historic mine, along with information on the commodity. **8.5" X 11", 88 ppgs, Retail Price: $9.99**

Metal Mines of Washington - Unavailable since 1921, this important publication was originally published by the Washington Geologic Survey and has been unavailable for nearly ninety years. Widely considered a masterpiece on the Washington Mining Industry, "Metal Mines of Washington" sheds light on the important details of Washington's early mining years. Featured are details on hundreds of gold, silver, copper, lead and other metallic mineral mines. Included are hard to find details on the mineral resources of this state, as well as the locations of historic mines. Lavishly illustrated with maps and historic photos and complete with a glossary to explain any technical terms found in the text, this is one of the most important works on mining in the State of Washington. No prospector or miner should be without it if they are interested in mining in Washington. **8.5" X 11", 396 ppgs, Retail Price: $24.99**

Gem Stones In Washington - Unavailable since 1949, this important publication was originally published by the Washington Geologic Survey and has been unavailable since first published. Included are details on where to find naturally occurring gem stones in the State of Washington, including quartz crystal, amethyst, smoky quartz, milky quartz, agates, bloodstone, carnelian, chert, flint, jasper, onyx, petrified wood, opal, fire opal, hyalite and others. **8.5" X 11", 54 ppgs, Retail Price: $8.99**

The Covada Mining District of Washington - Unavailable since 1913, this important publication was originally published by the Washington Geologic Survey and has been unavailable for a century. Topics include the geology, rock formations and the formation of ore deposits in this important mining area of Washington State. Also included are hard to find details on the geology, history and locations of dozens of mines in the area. Some of the mines featured include the Admiral, Advance, Algonkian, Big Bug, Big Chief, Big Joker, Black Hawk, Black Tail, Black Thorn, Captain, Cherokee Strip, Colorado, Dan Patch, Dead Shot, Etta, Good Ore, Greasy Run, Great Scott, Idora, IXL, Jay Bird, Kentucky Bell, King Solomon, Laurel, Laura S, Little Jay, Meteor, Neglected, Northern Light, Old Nell, Plymouth Rock, Polaris, Quandary, Reserve, Shoo Fly, Silver Plume, Three Pines, Vernie, White Rose and dozens of others. **8.5" X 11", 114 ppgs, Retail Price: $10.99**

The Index Mining District of Washington - Unavailable since 1912, this important publication was originally published by the Washington Geologic Survey and has been unavailable for a century. Topics include the geology, rock formations and the formation of ore deposits in this important mining area of Washington State. Also included are hard to find details on the geology, history and locations of dozens of mines in the area. Some of the mines featured include the Sunset, Non-Pareil, Ethel Consolidated, Kittaning, Merchant, Homestead, Co-operative, Lost Creek, Uncle Sam, Calumet, Florence-Rae, Bitter Creek, Index Peacock, Gunn Peak, Helena, North Star, Buckeye. Copper Bell, Red Cross and others. **8.5" X 11", 114 ppgs, Retail Price: $11.99**

Mining & Mineral Resources of Stevens County Washington - Unavailable since 1920, this important publication was originally published by the Washington Geologic Survey and has been unavailable for a century. Topics include the geology, rock formations and the formation of ore deposits in these important mining areas of Washington State. Also included are hard to find details on the geology, history and locations of hundreds of mines in the area. **8.5" X 11", 372 ppgs, Retail Price: $24.99**

The Mines and Geology of the Loomis Quadrangle Okanogan County, Washington - Unavailable since 1972, this important publication was originally published by the Washington Geologic Survey and has been unavailable for a century. Topics include the geology, rock formations and the formation of ore deposits in this important mining area of Washington State. Also included are hard to find details on the geology, history and locations of dozens of gold, copper, silver and other mines in the area. **8.5" X 11", 150 ppgs, Retail Price: $12.99**

The Conconully Mining District of Okanogan County Washington - Unavailable since 1973, this important publication was originally published by the Washington Geologic Survey and has been unavailable for a century. Topics include the geology, rock formations and the formation of ore deposits in this important mining area of Washington State, which also includes Salmon Creek, Blue Lake and Galena. Also included are hard to find details on the geology, mining history and locations of dozens of mines in the area. Some of the mines include Arlington, Fourth of July, Sonny Boy, First Thought, Last Chance, War Eagle-Peacock, Wheeler, Mohawk, Lone Star, Woo Loo Moo Loo, Keystone, Hughes, Plant-Callahan, Johnny Boy, Leuena, Gubser, John Arthur, Tough Nut, Homestake, Key and many others **8.5" X 11", 68 ppgs, Retail Price: $8.99**

Wyoming Mining Books

Mining in the Laramie Basin of Wyoming - Unavailable since 1909, this publication was originally compiled by the United States Department of Interior. Also included are insights into the mineralization and other characteristics of this important mining region, especially in regards to coal, limestone, gypsum, bentonite clay, cement, sand, clay and copper. **8.5" X 11", 104 ppgs, Retail Price: $11.99**

New Mexico Mining Books

The Mogollon Mining District of New Mexico - Unavailable since 1927, this important publication was originally published by the US Department of Interior and has been unavailable for 80 years. Topics include the geology, rock formations and the formation of ore deposits in this important mining area in New Mexico. Of particular focus is information on the history and production of the ore deposits in this area, their form and structure, vein filling, their paragenesis, origins and ore shoots, as well as oxidation and supergene enrichment. Also included are hard to find details, including the descriptions and locations of numerous gold, silver and other types of mines, including the Eureka, Pacific, South Alpine, Great Western, Enterprise, Buffalo, Mountain View, Floride, Gold Dust, Last Chance, Deadwood, Confidence, Maud S., Deep Down, Little Fanney, Trilby, Johnson, Alberta, Comet, Golden Eagle, Cooney, Queen, the Iron Crown, Eberle, Clifton, Andrew Jackson mine, Mascot and others. **8.5" X 11", 144 ppgs, Retail Price: $12.99**

The Percha Mining District of Kingston New Mexico - Unavailable since 1883, this important publication was originally published by the Kingston Tribune and has been unavailable for over one hundred and thirty five years. Having been written during the earliest years of gold and silver mining in the Percha Mining District, unlike other books on the subject, this work offers the unique perspective of having actually been written while the early mining history of this area was still being made. In fact, the work was written so early in the development of this area that many of the notable mines in the Percha District were less than a few years old and were still being operated by their original discoverers with the same enthusiasm as when they were first located. Included are hard to find details on the very earliest gold and silver mines of this important mining district near Kingston in Sierra County, New Mexico. **8.5" X 11", 68 ppgs, Retail Price: $9.99**

East Coast Mining Books

<u>The Gold Fields of the Southern Appalachians</u> - Unavailable since 1895, this important publication was originally published by the US Department of Interior and has been unavailable for nearly 120 years. Topics include the geology, rock formations and the formation of ore deposits in this important mining area of the American South. Of particular focus is information on the history and statistics of the ore deposits in this area, their form and structure and veins. Also included are details on the placer gold deposits of the region. The gold fields of the Georgian Belt, Carolinian Belt and the South Mountain Mining District of North Carolina are all treated in descriptive detail. Included are hard to find details, including the descriptions and locations of numerous gold mines in Georgia, North Carolina and elsewhere in the American South. Also included are details on the gold belts of the British Maritime Provinces and the Green Mountains. **8.5" X 11", 104 ppgs, Retail Price: $9.99**

Gold Rush Tales Series

<u>**Millions in Siskiyou County Gold**</u> - In this first volume of the "Gold Rush Tales" series, leading mining historian and editor Kerby Jackson, introduces us to the story of how millions of dollars worth of gold was discovered in Siskiyou County during the California Gold Rush. Lavishly illustrated with photos from the 19th Century, this hard to find information was first published in 1897 and sheds important light onto the gold rush era in Siskiyou County, California and the experiences of the men who dug for the gold and actually found it. **8.5" X 11", 82 ppgs, Retail Price: $9.99**

<u>**The California Rand in the Days of '49**</u> - In this second volume of the "Gold Rush Tales" series, leading mining historian and editor Kerby Jackson, introduces us to four tales from the California Gold Rush. Lavishly illustrated with photos from the 19th Century, this hard to find information was first published in 1890's and includes the stories of "California's Rand", details about Chinese miners, how one early miner named Baker struck it rich and also the story of Alphonzo Bowers, who invented the first hydraulic gold dredge. **8.5" X 11", 54 ppgs, Retail Price: $9.99**

More Mining Books

<u>**Prospecting and Developing A Small Mine**</u> - Topics covered include the classification of varying ores, how to take a proper ore sample, the proper reduction of ore samples, alluvial sampling, how to understand geology as it is applied to prospecting and mining, prospecting procedures, methods of ore treatment, the application of drilling and blasting in a small mine and other topics that the small scale miner will find of benefit. **8.5" X 11", 112 ppgs, Retail Price: $11.99**

<u>**Timbering For Small Underground Mines**</u> - Topics covered include the selection of caps and posts, the treatment of mine timbers, how to install mine timbers, repairing damaged timbers, use of drift supports, headboards, squeeze sets, ore chute construction, mine cribbing, square set timbering methods, the use of steel and concrete sets and other topics that the small underground miner will find of benefit. This volume also includes twenty eight illustrations depicting the proper construction of mine timbering and support systems that greatly enhance the practical usability of the information contained in this small book. **8.5" X 11", 88 ppgs. Retail Price: $10.99**

<u>**Timbering and Mining**</u> - A classic mining publication on Hard Rock Mining by W.H. Storms. Unavailable since 1909, this rare publication provides an in depth look at American methods of underground mine timbering and mining methods. Topics include the selection and preservation of mine timbers, drifting and drift sets, driving in running ground, structural steel in mine workings, timbering drifts in gravel mines, timbering methods for driving shafts, positioning drill holes in shafts, timbering stations at shafts, drainage, mining large ore bodies by means of open cuts or by the "Glory Hole" system, stoping out ore in flat or low lying veins, use of the "Caving System", stoping in swelling ground, how to stope out large ore bodies, Square Set timbering on the Comstock and its modifications by California miners, the construction of ore chutes, stoping ore bodies by use of the "Block System", how to work dangerous ground, information on the "Delprat System" of stoping without mine timbers, construction and use of headframes and much more. This volume provides a reference into not only practical methods of mining and timbering that may be employed in narrow vein mining by small miners today, but also rare insights into how mines were being worked at the turn of the 19th Century. **8.5" X 11", 288 ppgs. Retail Price: $24.99**

A Study of Ore Deposits For The Practical Miner - Mining historian Kerby Jackson introduces us to a classic mining publication on ore deposits by J.P. Wallace. First published in 1908, it has been unavailable for over a century. Included are important insights into the properties of minerals and their identification, on the occurrence and origin of gold, on gold alloys, insights into gold bearing sulfides such as pyrites and arsenopyrites, on gold bearing vanadium, gold and silver tellurides, lead and mercury tellurides, on silver ores, platinum and iridium, mercury ores, copper ores, lead ores, zinc ores, iron ores, chromium ores, manganese ores, nickel ores, tin ores, tungsten ores and others. Also included are facts regarding rock forming minerals, their composition and occurrences, on igneous, sedimentary, metamorphic and intrusive rocks, as well as how they are geologically disturbed by dikes, flows and faults, as well as the effects of these geologic actions and why they are important to the miner. Written specifically with the common miner and prospector in mind, the book will help to unlock the earth's hidden wealth for you and is written in a simple and concise language that anyone can understand. **8.5" X 11", 366 ppgs. Retail Price: $24.99**

Mine Drainage - Unavailable since 1896, this rare publication provides an in depth look at American methods of underground mine drainage and mining pump systems. This volume provides a reference into not only practical methods of mining drainage that may be employed in narrow vein mining by small miners today, but also rare insights into how mines were being worked at the turn of the 19th Century. **8.5" X 11", 218 ppgs. Retail Price: $24.99**

Fire Assaying Gold, Silver and Lead Ores - Unavailable since 1907, this important publication was originally published by the Mining and Scientific Press and was designed to introduce miners and prospectors of gold, silver and lead to the art of fire assaying. Topics include the fire assaying of ores and products containing gold, silver and lead; the sampling and preparation of ore for an assay; care of the assay office, assay furnaces; crucibles and scorifiers; assay balances; metallic ores; scorification assays; cupelling; parting' crucible assays, the roasting of ores and more. This classic provides a time honored method of assaying put forward in a clear, concise and easy to understand language that will make it a benefit to even beginners. **8.5" X 11", 96 ppgs. Retail Price: $11.99**

Methods of Mine Timbering - Originally published in 1896, this important publication on mining engineering has not been available for nearly a century. Included are rare insights into historical methods of timbering structural support that were used in underground metal mines during the California that still have a practical application for the small scale hardrock miner of today. **8.5" X 11", 94 ppgs. Retail Price: $10.99**

The Enrichment of Copper Sulfide Ores - First published in 1913, it has been unavailable for over a century. Topics include the definition and types of ore enrichment, the oxidation of copper ores, the precipitation of metallic sulfides. Also included are the results of dozens of lab experiments pertaining to the enrichment of sulfide ores that will be of interest to the practical hard rock mine operator in his efforts to release the metallic bounty from his mine's ore. **8.5" X 11", 92 ppgs. Retail Price: $9.99**

A Study of Magmatic Sulfide Ores - Unavailable since 1914, this rare publication provides an in depth look at magmatic sulfide ores. Some of the topics included are the definition and classification of magmatic ores, descriptions of some magmatic sulfide ore deposits known at the time of publication including copper and nickel bearing pyrrohitic ore bodies, chalcopyrite-bornite deposits, pyritic deposits, magnetite-ileminite deposits, chromite deposits and magmatic iron ore deposits. Also included are details on how to recognize these types of ore deposits while prospecting for valuable hardrock minerals. **8.5" X 11", 138 ppgs. Retail Price: $11.99**

The Cyanide Process of Gold Recovery - Unavailable since 1894 and released under the name "The Cyanide Process: Its Practical Application and Economical Results", this rare publication provides an in depth look at the early use of cyanide leaching for gold recovery from hardrock mine ores. This volume provides a reference into the early development and use of cyanide leaching to recover gold. **8.5" X 11", 162 ppgs. Retail Price: $14.99**

California Gold Milling Practices - Unavailable since 1895 and released under the name "California Gold Practices", this rare publication provides an in depth look at early methods of milling used to reduce gold ores in California during the late 19th century. This volume provides a reference into the early development and use of milling equipment during the earliest years of the California Gold Rush up to the age of the Industrial Revolution. Much of the information still applies today and will be of use to small scale miners engaging in hardrock mining. **8.5" X 11", 104 ppgs. Retail Price: $10.99**

Leaching Gold and Silver Ores With The Plattner and Kiss Processes - Mining historian Kerby Jackson introduces us to a classic mining publication on the evaluation and examination of mines and prospects by C.H. Aaron. First published in 1881, it has been unavailable for over a century and sheds important light on the leaching of gold and silver ores with the Plattner and Kiss processes. **8.5" X 11", 204 ppgs. Retail Price: $15.99**

The Metallurgy of Lead and the Desilverization of Base Bullion - First published in 1896, it has been unavailable for over a century and sheds important light on the the recovery of silver from lead based ores. Some of the topics include the properties of lead and some of its compounds, lead ores such as galenite, anglesite, cerussite and others, the distribution of lead ores throughout the United States and the sampling and assaying of lead ores. Also covered is the metallurgical treatment of lead ores, as well as the desilverization of lead by the Pattinson Process and the Parkes Process. Hofman's text has long been considered one of the most important early works on the recovery of silver from lead based ores. 8.5" X 11", 452 ppgs. Retail Price: $29.99

Ore Sampling For Small Scale Miners - First published in 1916, it has been unavailable for over a century and sheds important light on historic methods of ore sampling in hardrock mines. Topics include how to take correct ore samples and the conditions that affect sampling, such as their subdivision and uniformity. Particular detail is given to methods of hand sampling ore bodies by grab sample, pipe sample and coning, as well as sampling by mechanical methods. Also given are insights into the screening, drying and grinding processes to achieve the most consistent sample results and much more. 8.5" X 11", 124 ppgs. Retail Price: $12.99

The Extraction of Silver, Copper and Tin from Ores - First published in 1896, it has been unavailable for over a century and sheds important light on how historic miners recovered silver, copper and tin from their mining operations. The book is split into three sections, including a discussion on the Lixiviation of Silver Ores, the mining and treatment of copper ores as practiced at Tharsis, Spain and the smelting of tin as it was practiced by metallurgists at Pulo Brani, Singapore. Also included is an overview and analysis of these historic metal recovery methods that will be of benefit to those interested in the extraction of silver, copper and tin from small mines. 8.5" X 11", 118 ppgs. Retail Price: $14.99

The Roasting of Gold and Silver Ores - First published in 1880, it has been unavailable for over a century and sheds important light on how historic miners recovered gold and silver rom their mining operations. Topics include details on the most important silver and free milling gold ores, methods of desulphurization of ores, methods of deoxidation, the chlorination of ores, methods and details on roasting gold and silver ores, notes on furnaces and more. Also included are details on numerous methods of gold and silver recovery, including the Ottokar Hofman's Process, the Patera Process, Kiss Process, Augustin Process, Ziervogel Process and others. 8.5" X 11", 178 ppgs. Retail Price: $19.99

The Examination of Mines and Prospects - First published in 1912, it has been unavailable for over a century and sheds important light on how to examine and evaluate hardrock mines, prospects and lode mining claims. Sections include Mining Examinations, Structural Geology, Structural Features of Ore Deposits, Primary Ores and their Distribution, Types of Primary Ore Deposits, Primary Ore Shoots, The Primary Alteration of Wall Rocks, Alterations by Surface Agencies, Residual Ores and their Distribution, Secondary Ores and Ore Shoots and Vein Outcrops. This hard to find information is a must for those who are interested in owning a mine or who already own a lode mining claim and wish to succeed at quartz mining. 8.5" X 11", 250 ppgs. Retail Price: $19.99

Garnets: Their Mining, Milling and Utilization - First published in 1925, it has been unavailable since those days and sheds important light on the mining, milling and utilization of garnets. Included are details on the characteristics of garnets, where they are found and how they were mined. 78 ppgs, 10.99

Gemstones and Precious Stones of North America - Leading mining historian Kerby Jackson introduces us to a classic mining publication on the gems and precious stones of the United States, Canada and mexico. First published in 1890, it has been unavailable since those days and sheds important light on the gems and precious stones that may be found in North America. Included are chapters on diamonds, corundum, sapphire, ruby, topaz, emerald, disapore, spinel, turquoise, tourmaline, garnets, beyrl, peridot, zircon, quartz crystals, feldspars, pearls and many others. Included are details on where these gems and precious stones may be found throughout North America, as well as their characteristics. 360 ppgs, 24.99

Mining Camps and Mining Districts - First released in 1885 by Charles Howard Shinn under the title "Mining Camps: A Study in American Frontier Government", this publication offers a unique look at how early gold miners established their own forms of representative government during the California Gold Rush. Drawing on the the early mining codes of mideviel German miners in the Harz Mountains, on the mining customs of the Cornish tin miners and early Spanish mining laws introduced into California, the miners established the first governments in the American West. 340 ppgs, 24.99

BLM Field Handbook for Mineral Examiners - Leading mining historian Kerby Jackson introduces us to a classic mining publication on mine evaluation. First published in 1962, this work sheds important light on the techniques of BLM Mineral Examiners to perform validity on mining claims. 132 ppgs, 10.99

<u>**Six Months In The Gold Mines During The California Gold Rush**</u> - Unavailable since 1850, this important work is a first hand account of one "49'ers" personal experience during the great California Gold Rush, shedding important light on one of the most exciting periods in the history of not only California, but also the world. Compiled from journals written between 1847 and 1849 by E. Gould Buffum, a native of New York, "Six Months In The Gold Mines During The California Gold Rush" offers a rare look into the day to day lives of the people who came to California to work in her gold mines when the state was still a great frontier. **8.5" X 11", 290 ppgs. Retail Price: $19.99**

<u>**The Discovery of Gold in Australia**</u> - First published in 1852, it has been unavailable since those days and sheds important light on Australia's gold mining history. Included are rare communications between British agents and the British Crown when gold was first discovered in Australia in 1851. This rare text contains hard to find details on Australia's first mining camps and Britain's early attempts to provide for the orderly regulation of gold mines in that part of the world. Also of interest are hard to find extracts of articles that appeared in the early colonial newspapers that did their best to report on Australia's gold rush as it took place.
102 ppgs, 10.99

www.ingramcontent.com/pod-product-compliance
Lightning Source LLC
Chambersburg PA
CBHW082303200526
45168CB00017B/2751